PANDEMIC SOCIETIES

A Critical Public Health Perspective

Alan Petersen

BRISTOL
UNIVERSITY
PRESS

First published in Great Britain in 2024 by

Bristol University Press
University of Bristol
1–9 Old Park Hill
Bristol
BS2 8BB
UK
t: +44 (0)117 374 6645
e: bup-info@bristol.ac.uk

Details of international sales and distribution partners are available at bristoluniversitypress.co.uk

British Library Cataloguing in Publication Data
A catalogue record for this book is available from the British Library

ISBN 978-1-5292-2036-0 hardcover
ISBN 978-1-5292-2037-7 paperback
ISBN 978-1-5292-2038-4 ePub
ISBN 978-1-5292-2039-1 ePdf

Cover design: Andrew Corbett
Front cover image: Getty Images/DrAfter123
Bristol University Press uses environmentally responsible print partners.
Printed and bound in Great Britain by CPI Group (UK) Ltd, Croydon, CR0 4YY

FSC
www.fsc.org
MIX
Paper | Supporting
responsible forestry
FSC® C013604

Contents

Acknowledgements — iv

1	Pandemics as Socio-Political Phenomena	1
2	The Politics of Framing a Pandemic Crisis	25
3	Modelling Pandemics	50
4	Pandemic Crisis and Inequalities	71
5	Pandemic Crisis and Technological Change	86
6	Future Pandemic Societies	108

References — 130

Index — 162

Acknowledgements

This book has had a long gestation, being conceived in 2020 during a series of COVID-19 lockdowns in that year. I am grateful to Paul Stevens, Senior Commissioning Editor at Bristol University Press, who has supported this project from the outset and accommodated my requests for extensions. Ellen Mitchell, Emma Cook, Georgina Bolwell, Kathryn King, and Freya Trand at Bristol University Press have also assisted in various ways, and I thank them. I would also like to thank the three reviewers who saw merit in the proposal and provided useful suggestions and encouragement, and the two readers of the manuscript who offered valuable guidance for revision. I have been fortunate in receiving a period of teaching relief from Monash University through the Outside Studies programme, which allowed time for writing, for which I am grateful. Finally, I am deeply indebted to Ros Porter who has patiently listened to me discussing my ideas and has offered her love and support over the years. I would like to dedicate this book to the many people who have suffered illness or lost their lives or otherwise been detrimentally impacted by the COVID-19 pandemic.

1

Pandemics as
Socio-Political Phenomena

According to many commentators, we are now living in an age of recurrent epidemics and pandemics. Since 2000, there has been a resurgence of infectious disease outbreaks including SARS, H1N1, Ebola, Zika, MERS, and COVID-19. Each infectious outbreak is seen to present new challenges but has been met with broadly similar responses, albeit implemented with varying degrees of urgency; namely, efforts to mitigate and manage the risks; the search for new diagnostic tools; the launch of new programmes of research, and a hopeful wait for potentially life-saving vaccines and/or treatments. Yet, while these responses were evident during COVID-19, this event seems distinctive in various respects.

The World Health Organization's declaration of a worldwide pandemic on 11 March 2020 – widely described as an 'unprecedented crisis' – resulted in measures that suddenly and profoundly disrupted economies, societies, and individuals' lives around the world. It also inaugurated a host of new social practices such as 'work-from-home' or 'hybrid working' and the rapid development and application of various new technologies of risk management. Official responses to the COVID-19 pandemic called for specific kinds of citizens: ones constantly alert to the prospect of infection and disease and cognisant of their responsibilities to manage their own relationship to risk and the risk they pose to others. However, it was often unclear what it meant to 'manage risk' in the context of a rapidly evolving disease surrounded with many uncertainties and unknowns and subject to constantly shifting and often conflicting information and advice communicated by various authorities, credentialed experts, news media, and online forums.

In this book I offer a critical public health perspective on the dimensions, dynamics, and implications of emerging pandemic societies, drawing on insights gained from analysing the constructions of and responses to COVID-19, as well as research on earlier epidemics and pandemics. I consider the role that

1

science and technology have played during COVID-19, and the lessons that may be drawn from this for public health in the future. In undertaking this analysis, I utilize ideas from various strands of sociology, especially work on frames and framing and on crises, and from science and technology studies and the sociology of emotions. Although, historically, sociologists have paid relatively little attention to the study of epidemics and pandemics and the field is still nascent (Dingwall, 2023), sociology offers a useful toolkit of ideas for making sense of the constructions, disruptions, and contestations that surround such events. This book, then, joins a relatively small but rapidly growing literature on the topic, with increased interest clearly sparked by the COVID-19 pandemic.

As sociologists well recognize – but policy makers and public health and medical experts seemingly less so (as responses to COVID-19 suggest) – pandemics are socio-political phenomena as well as biological phenomena and thus need to be understood in relation to the specific conditions that give rise to them and shape the responses (see, for example, Dingwall et al, 2013). They are subject to interpretation as to their origins, magnitude, implications, and the required interventions. As COVID-19 has shown, different groups with different agendas compete to define a pandemic's significance and there is much at stake in the ability to successfully 'frame' such an event, especially during its early stages before related policies and programmes become settled or 'locked in' and difficult to change.

I believe that a critical public health perspective on the COVID-19 pandemic will not only advance understanding of the dynamics, dimensions, and implications of this event but can assist in the effort to develop alternative, potentially less personally and socially harmful responses to pandemics in the future. In short, the crisis conditions of COVID-19 offer the prospect of social renewal; of taking steps to create societies that are more equal, less polarized, and more attune to the diversity of people's views and experiences, which are shaped by many factors including gender, age, ethnicity, nationality, sexuality, and physical and mental ability. I discuss how sociologists and other critical scholars may contribute to advancing such an agenda in the concluding chapter.

A major question arising from my analysis of the evidence on the COVID-19 pandemic thus far is: could alternative interpretations of and responses to this event have had different outcomes – perhaps ones less damaging than those experienced? The personal and social disruptions and harms exacted by pandemic measures, especially long lockdowns, including loss of employment or closure of businesses, impacts on young people's education, heightened anxieties, experiences of isolation and loneliness, self-reported increases in alcohol and drug abuse and mental illness, domestic violence, and the increased incidence of child sexual exploitation and other online harms (UNICEF, 2020; Salter and Wong, 2021), are incalculable.

Official responses have often served to marginalize and stigmatize people, including those unwilling or unable to conduct themselves as ascribed 'responsible' citizens or who have questioned or resisted certain mitigation measures such as vaccinations, mandated masks, and social distancing for ideological, cultural, religious, financial, ethical, safety, historical, or other reasons (Parmet et al, 2005; Olick et al, 2021; De Wit et al, 2023). The longer-term economic and social costs resulting from adopted measures, including the cascading impacts on economies, social stability and international security, delayed diagnoses and treatment of diseases, and effects on productivity, are also immeasurable and it may be many years before the full ramifications are known.

What *is* clear from research in the three years following the declaration of the emergency is that the short-to-medium impacts of the disease itself and of measures to control the spread of infection have been unevenly experienced across populations. The COVID-19 pandemic laid bare many inequalities – some already existing and some resulting from or reinforced by related measures – between relatively resource-rich and resource-poor countries and between those of different genders, ages, ethnic/racial groups, and socio-economic backgrounds (Chapter 4).

In this chapter and the chapters that follow, I examine some of the substantial research evidence published to date which is produced by governments, expert panels, international organizations such as the World Health Organization (WHO), the United Nations (UN), and the World Bank, university researchers, and reports produced by think-tanks and non-governmental organizations. Before proceeding further, however, I should explain the context and aims of my book and how it differs from others on COVID-19 published to date, after which I provide an overview of my guiding concepts, assumptions, and overall argument.

Covidization of research

I developed the proposal for my book in the first year of COVID-19 when people around the world were trying to make sense of the pandemic, its dimensions, and implications. It was a period of great fear and uncertainty, and much debate about how best to protect people from the risk of infection, illness, and death, which was the priority at the time. It was then unclear how long it would last and how exactly it would impact upon economies, societies, and individuals. The question of how the media was or should be reporting the issues was also extensively debated, especially with the deluge of so-called misinformation and rumours rapidly circulating online and the emergence of the 'home-based expert' (Kirkland and Fang, 2023). Many scholars and practitioners in different fields then and since have been grappling with these issues, which led to the generation of a huge quantity of research and writing spanning many specialisms, along with much media commentary.

In the first year of the pandemic, journals were confronted with a deluge of submissions on COVID-19, with one writer commenting that the 'covidization of research' was occurring with such urgency that it threatened 'careful, deliberate discourse or scholarship' (Pai, 2020). Editors of journals in the fields of public health and social science received significant increases in publication submissions during COVID-19. The editors of *Critical Public Health* (for which I served as editorial board member) wondered whether the rush to publish was leading to 'premature evaluation' and 'crystallising inequalities in scholarly knowledge production (gendered and otherwise) in much the same ways that it is serving to highlight inequalities in other areas' (Bell and Green, 2020: 380).

By the end of October 2020, 125,000 articles on COVID-19 had been published, of which 25 per cent were preprints (Fraser et al, 2021). Almost a year later, this number had more than quadrupled. In an article examining open access, published in September 2021, Jeffrey Brainard noted that 530,000 new articles on COVID-19 had appeared, published either by journals or as preprints, which, he claims, 'fed the largest 1-year increase in all scholarly articles, and the largest annual total ever' (Brainard, 2021). He commented that one of the leading servers, *medRxiv*, which was founded as recently as 2019, 'posted about 200 preprints on all topics in January 2020; by May 2020, the monthly tally had swelled to more than 2000, about three-quarters of them about the pandemic virus' (Brainard, 2021; see also Asai, 2023).

In addition to this proliferation of scientific studies, many articles, monographs, and edited collections have explored the social and political implications of the COVID-19 pandemic. These have been contributed by sociologists and other social scientists, philosophers, media studies scholars, public policy analysts, and other specialists (for example, Delanty, 2021; Fassin and Fourcade, 2021; Ryan, 2021; Bringel and Pleyers, 2022; Brown and Zinn, 2022; Gadarian et al, 2022; Lupton, 2022; Lyon, 2022; Feierstein, 2023; Lewis et al, 2023; Moyo and Ndlovu-Gastsheni, 2024). In the first two years of the pandemic, many books and edited collections were published in the form of 'rapid responses' or 'accelerated publications', in the effort to get work out quickly to influence thinking, research, and policy.

While these publications provided many valuable insights into the pandemic and its impacts, most are limited by their focus on relatively short intervals of time – typically within the first and/or second year of the pandemic (2020 and/or 2021), when its impacts were only just beginning to be grasped. The COVID-19 research agenda has been guided in part by the priorities of national research councils, charities like the UK's Wellcome Trust, and universities that funded a great deal of work on COVID-19 soon after its onset. Researchers' use of relatively short time frames makes it difficult to know whether the documented trends will endure or whether the implications of practices or programmes are as far reaching as envisaged at the time of the research. All studies are limited by their time frames and their findings are always subject to reinterpretation and re-evaluation in the future. However, I suggest that the research agenda on COVID-19 has been profoundly shaped by the way the pandemic was *initially framed*, which, as mentioned, generated great uncertainty and fear.

For example, a substantial body of literature produced during COVID-19 examines concepts such as solidary, civility, and compassion which were brought into sharp focus during the first two years of the pandemic when these were emphasized in official discourses (for example, World Economic Forum, 2020) but often found to be in short supply (for example, Bonotti and Zech, 2021; Gandenberger et al, 2023; Kieslich et al, 2023). The COVID-19 crisis also generated a proliferation of pandemic stories as people began to share their experiences of lockdowns and their restrictions, of hospitalizations, of front-line health work, of loneliness, and of Long COVID (Zhang et al, 2021; Ireson et al, 2022; Hayden et al, 2023; Mehta et al, 2023). The event was highly disruptive and impacted upon people's lives in many ways, and so it is hardly surprisingly that individuals wanted to share their stories and that researchers wanted to document them. Yet, this work, much published as journal articles or chapters in edited collections – often published as 'rapid response' pieces or special issues on COVID-19 (of which there have been many) – offers only snapshots of a complex, rapidly evolving event whose full implications will not be known for many years. For researchers, it is difficult to analyse an event such as COVID-19 when one is in its midst and a longer time frame is needed to properly evaluate its significance.

My aim in writing this book is to offer a wider lens on events than that offered by other researchers to date, to encourage reflection on what might be learnt from analysing the framing of 'the COVID-19 crisis' and the subsequent responses about the values, priorities, and dynamics of societies grappling with recurrent pandemics – or what I call pandemic societies. Given that the pandemic was officially declared as being over by the WHO on 5 May 2023 (World Health Organization, 2023), I have been able to examine events as they have unfolded, beginning with WHO's declaration of the pandemic on 11 March 2020 to its conclusion. No other sociological study, to my knowledge, has examined COVID-19 over its duration from its beginning to its end. I am especially interested in advancing understanding of the role played by science and technology in mediating the crisis and in efforts to mitigate and manage related risks during this period.

Science and technology figured prominently in pandemic control efforts but, as became increasingly evident over time, could not always be relied on to provide the certainty that policy makers and citizens sought. As I explain (especially in Chapter 2), risk governance, which relies on science and technology for prediction and control, was shown to be limited in a context of great uncertainty, including regarding human responses to pandemic measures.

Assumptions, concepts, and argument

In the following paragraphs, I elaborate my guiding assumptions, concepts, and argument, and conclude by summarizing the content of the remaining chapters. I begin by examining the concepts of frame and framing and their utility in understanding the dimensions and dynamics of a pandemic crisis like COVID-19. I then discuss the critical role ascribed to science and technology in mitigating and managing pandemic risks, and the significance of popular cultural portrayals in shaping responses, including apportioning blame and responsibility for contagious outbreaks. Next, I examine the dimensions and dynamics of crises, drawing on sociological work that helps shed light on their social definition and characterization. I also discuss the role that emotional regimes play in sustaining a crisis and shaping the responses – the dimensions of which I elaborate in Chapter 2. Finally, I examine the limitations and failures of so-called risk governance, which became evident as the pandemic unfolded, especially so in the practices of epidemiological modelling – the focus of Chapter 3.

Frames and framing

The concepts of frame and framing I believe are useful in exploring the dimensions and dynamics of pandemics: how they are understood and responded to, the range of measures adopted, and the subsequent implications. As it was conceived by sociologist Erving Goffman in his seminal work on frame analysis, a 'frame' refers to the ways experience is organized by 'definitions of the situation' (1986; orig. 1974: 10–11). Frames provide 'schemata of interpretation', or models of what should demand our attention (1986: 21). In media studies, frame analysis (or framing analysis) has been used by scholars to highlight how competing claims-makers seek to portray issues in particular ways by emphasizing certain aspects and ignoring others, thereby limiting policy debate on issues (Miller and Riechert, 2000: 46–47). Framing is assisted by use of a particular language, images, stereotypes, and other rhetorical techniques that assist readers/ audiences to understand a situation or activity. In analysing how pandemics are framed or defined, I suggest, important insights can be gained into a society's values and priorities as well as tensions and contradictions in the social order. An analysis of frames and framing can also help explain why pandemic measures like those implemented to help control the spread of SARS–CoV–2 infection may fail to fulfil their promise and/or have unforeseen consequences.

Science and technology have played a central role in mediating the pandemic and efforts to mitigate and manage its risks. The rapid production of research and access to findings, and the introduction of new technologies, including vaccines, rapid antigen self-tests (RATs) (also variously called lateral flow tests, at-home tests or just rapid tests or antigen tests), and AI-enabled devices of various kinds, has been phenomenal, and the implications for conceptions of science and society are far reaching. Science and technology, and related expertise, figure prominently in my discussion since they have profoundly shaped public understandings of and reactions to COVID-19 through their applications in diverse domains including news and other media, epidemiological modelling, the creation of online dashboards, contact tracing, vaccine development, diagnostic innovations, and anti-viral treatments.

Clearly, many hopes have attached to science and technology, for their potential to quickly advance understanding of and help prevent disease (for example, open online access to research findings via preprint articles, the sharing of data in real time), predict and track the transmission of infection, develop strategies of risk management, and evaluate the effectiveness of different interventions. In the book I ask: have these hopes been misplaced? Is too much expected of science and technology? Might the search for technological solutions have diverted attention from other potential responses?

Within official discourses, the COVID-19 pandemic was framed as a 'crisis' or 'a state of exception', which is a particular definition of the situation that enables authorities to extend powers that they would not ordinarily exercise (Agamben, 2005). These powers led to suspending citizens' existing rights including freedom of movement and association, and accorded considerable authority to the prescribed 'risk managers', public health experts. One can find many historical examples of states of exception, including the extension of surveillance policies following the 9/11 attacks, and governments' use of extraordinary powers during periods of civil unrest, war, economic crisis, biosecurity risk, and events such as floods, fires, hurricanes, and nuclear disasters. These powers are generally justified on the grounds that they are necessary for the safety and protection of citizens. The question of whether this extension of powers is proportionate to the actual threat/s to citizens in any specific situation or oversteps acceptable standards by imposing sweeping and overly broad restrictions in a community is debatable, and during COVID-19 was subject to much discussion by human rights organizations and within the wider community (for example, Human Rights Watch, 2020; Mijatović, 2020; Council of Europe, 2023).

As UN Secretary-General António Guterres was reported in the *Guardian*, in February 2021, as saying, COVID-19 mitigation measures provided the context for a 'pandemic of human rights abuses', involving 'a global crackdown on opposition activists and human rights defenders, increased attacks on journalists and moves to curb free speech, censor the media, roll out invasive tracking apps and put in place extreme surveillance measures, many of which are likely to far outlast the pandemic' (Kelly and Pattisson, 2021). With emergencies in the past, citizens have sometimes resisted the extension of the new state powers, which was evident at different stages during the COVID-19 pandemic. Regardless of whether they can be justified, related measures will have consequences which in many if not most cases will not have been anticipated by those who called for and supported the extension of powers.

At different phases of COVID-19, different groups of actors sought to reframe the pandemic, including its origins, magnitude, implications, and the required responses. The news and other media play a crucial role in framing issues on matters of public interest, and this role was evident during the COVID-19 pandemic. In Australia, research has shown that people were accessing more news than usual via traditional (or legacy) media, especially television, during the early stages of COVID-19 (April 2020) in their search for credible sources of information on the pandemic (Park et al, 2020). Research on news consumption from late February to early March of 2020 found a similar pattern for the US and Europe (Casero-Ripollés, 2020). Analyses of news media framing reveals that the early definition of events is critical for how they are understood by audiences and for subsequent portrayals (for example, Petersen, 2002; Anderson et al, 2009).

I will explain how contestations over the framing of issues occurred at various stages in the pandemic's unfolding; for example, regarding the reality of the pandemic and the impact of various proposed or adopted measures (Chapter 2), and the utility of different models (Chapter 3). During its early phase, different stakeholders in different countries sought to frame COVID-19 either as an event of relatively little consequence (as some claimed, 'like the seasonal flu') or as a catastrophe that required a 'military-like' response and/or as a 'risk' event that could be 'managed' and 'mitigated' using certain technologies and techniques. As the pandemic progressed and it became evident that vaccines per se would not stem infections since new variants, such as Delta, and then Omicron, would continue to cause disease and death, many politicians and businesses began to argue that societies would need to 'live with the virus' and accept a level of disease and death that would likely have been unacceptable in the early phases of the pandemic. The stakes of framing have been high, for individual citizens, businesses, and policy makers, since, as history shows, once policies are implemented, they can be difficult to reverse.

The role of digital technologies

A major feature of the COVID-19 pandemic that distinguishes it from earlier pandemics is that it was the first to occur in the digital age characterized by the widespread use of social media and artificial intelligence (AI) that employ algorithms 'trained' on large datasets. Digital technologies have shaped the representations of the COVID-19 pandemic and the responses and how science was communicated among scientists and health authorities and between the latter two groups and the wider community. Various experts, including epidemiologists, virologists, and biostatisticians, employed technologies powered by AI to undertake their research and communicate their findings. Crisis conditions served to accelerate the development of AI, as Big Tech and other companies increased investment in innovations as many citizens became dependent on the internet during the lockdowns and other restrictions, as I discuss in Chapter 5.

During COVID-19, ordinary citizens also used technologies to learn about and exchange stories about the pandemic and, in some cases, challenge policies and accredited experts. Digital technologies provide a new means for enacting what I call 'bio-digital citizenship'; that is, the entwining of biological-based identities, rights and responsibilities, and digital practices (Petersen, 2019; Petersen et al, 2019). The rapid growth in the use of these technologies during the pandemic has served to accelerate 'data pooling' whereby data drawn from different sources are combined by technology companies to develop new innovations such as generative AI (Chapter 5). I will discuss these impacts of digital technologies later in this chapter and the chapters that follow.

The COVID-19 pandemic has offered many insights into how science, technology, and politics interact and shape representations of and responses to a crisis event. As a health sociologist whose work also spans science and technology studies (STS), critical public health, and the sociology of emotions, I am interested in uncovering the implications of extensive reliance on technologies for tackling a pandemic and how promise, hope, and fear shape views and actions. This book builds on other work in STS that explores the historical and socio-political conditions affecting the development of and citizens' engagements with science and technology. Much of this work explores how science 'facts' are constructed and contested by different groups, and how certain claims come to be viewed as credible or not. In addition to these questions, I am interested in understanding how a sense of urgency is created by the declaration of an emergency – a 'crisis' demanding decisive, internationally coordinated action – and why the responses may fall short of what is demanded by authorities.

While some governments took such action in response to the WHO's declaration of 'a public health event of international concern', by enacting national lockdowns, border closures, contact tracing, and other measures, to begin with, and then later mass vaccinations, others did not – because countries were ill equipped and ill prepared or did not have the resources required to do so (as has been the case with many countries in economically poorer parts of the world). In fact, the responses were often chaotic, inconsistent, and ever changing, and evidently had less to do with science per se than with national or local political factors (Chapter 2). However, to say this is not to suggest that science is apolitical or value neutral. To the contrary, science is inescapably political – a fact that was brought into sharp relief during COVID-19.

China's sudden change in direction in its pandemic strategy, from harsh lockdowns to open borders, following mass demonstrations in the latter months of 2022 vividly illustrated the workings of politics and power in pandemic responses, and the way science may be used selectively by authorities to justify their decisions, including changes in policies. However, while China provides a well-publicized case, many other national examples can be cited of how pandemic policies have reflected the workings of politics and power, including resistance to measures. An especially interesting example worth mentioning in this regard is Sweden, whose responses received considerable news coverage early in the pandemic.

In 2020, when the UK and most of Europe was in government-enforced lockdowns, following modelling suggesting that a massive number of deaths would occur under a worse-case scenario if no measures were taken (see Chapter 3), Sweden decided to keep schools, bars, restaurants, and shops open while encouraging people to work from home and socially distance, on the assumption that the country would achieve 'herd immunity', whereby most people would become infected and develop protection, thus preventing the spread of the virus. According to an assessment of 'the Swedish experiment', the 'state epidemiologist', Anders Tegnell, along with one of his predecessors in this function and his mentor, Johan Giesecke (who came out of retirement to serve as a consultant to the Public Health Agency (FME)), 'argued in favour of a piecemeal, evidence-based approach, a mitigation strategy that would imply living with the virus, as they believed that it would inevitably spread widely and could not be kept at bay with stop-and-go policies that would prove unacceptable in the long run' (Aucante, 2022: 2). As Aucante explains, lockdowns were seen as measures of last resort that had potentially adverse consequences and it would be 'difficult to decide when and how to exit from this state of closure and to avoid flare-ups absent any reliable treatment' (2022: 2).

Sweden's approach attracted much criticism, both at home and abroad, as it would amount to 'letting the virus run free and waiting for some herd immunity' (Aucante, 2022: 2). To be sure, citizens were advised to take other measures, such as washing their hands, social distancing, caps on social gatherings, and working remotely, when possible, and some restrictions were implemented such as the requirement that high schools and universities convert to distance education (advice that seemed somewhat at odds with the hoped-for herd immunity) (2022: 1, 3). But Sweden was generally considered an exceptional case in its approach in Europe and, indeed, internationally.

By June of 2020, however, it was reported that the country had one of the highest per capita rates of coronavirus deaths in the world and that a study conducted by the country's Public Health Agency had shown that 'the number of Swedes who had developed antibodies to the virus is smaller [6.1 per cent] than expected [40 per cent was predicted by Anders Tegnell, the country's chief epidemiologist], dashing hopes that herd immunity can be achieved' (News.com.au, 2020). Yet, faced with this alarming evidence, the chief epidemiologist argued that 'the country does not need to radically alter its policy' and was cited as saying ' "The strategy has never been to achieve a certain level of immunity. ... Our strategy has always been to keep the level of spread on a level that is so low that it does not affect society or healthcare in any catastrophic way and that has been achieved" ' (News.com. au, 2020); adding: ' "We have different levels of immunity on different parts of the population at this stage, from 4 percent to 5 percent to 20 percent to 25 percent" ' (News.com.au, 2020).

That is, faced with new 'facts' that were discordant with previous ones, Swedish authorities changed the narrative, to argue that it was never the government's intention to achieve a certain level of immunity. Later, Swedish researchers challenged the interpretation that non-pharmaceutical interventions (NPIs) and 'herd immunity' protection accounted for the rise and fall of COVID-19 epidemic waves in 'hard-hit locations where authorities have not enforced strict NPIs' (Robertson, 2021). (For a detailed discussion of Sweden's responses, including the failure of its experts to radically modify their approach and recommendations, even when natural immunity was not imminent, and the political, institutional, and policy context shaping its approach, see Aucante, 2022.)

The politics of science

Scientists tend to think of their work as being objective and apolitical – and thus devoid of values, beliefs, and metaphors – but considerable scholarship in philosophy and social studies of science shows that this is far from the case (for example, Agassi, 2003). The COVID-19 pandemic exposed the fallacy of a value-free or politically neutral science. Scientists' work is enmeshed in relations of power, and recognizing this has important implications in planning for future pandemics. In Chapter 3, I explore science-as-politics in action as manifest in modelling, which played a crucial role in pandemic responses during COVID-19. Both the language used to describe the COVID-19 pandemic and the methods used to study it are performative or prescriptive in suggesting how it *should* be understood. As sociologist John Law argues, methods help create rather than just describe realties, and are thus 'always political'. The use of scientific categories and labels both make visible and render invisible certain realities (Law, 2004).

One will never know what the consequences of the COVID-19 pandemic would have been had it been framed and responded to differently than how it was. One can only speculate. It is crucial to consider whether a different framing/s and related responses may have had different effects and perhaps ones less deleterious than those experienced during the pandemic. Many claims have been made about lives saved and illnesses prevented by the implementation of lockdowns and other measures. But these have generally been made by credentialed experts, policy makers, and other actors with a stake in those decisions and the associated imagined futures. Retrospective accounts are prone to (re)interpretation as vested interests seek to justify decisions made and to bring history into line with current views and practices and future visions. The benefits of various measures, such as lockdowns and mass testing and mass vaccination, have been both justified and questioned with reference to 'science', but the latter has been found wanting in addressing many questions and uncertainties raised by the rapidly evolving pandemic. Fundamental assumptions and approaches to governance, and specifically 'risk governance', have been brought into question, as I will explain.

My research leads me to question the premise that there is or could be general agreement on what constitutes a pandemic crisis, that such an event unfolds in a certain way with a clear beginning and end, and that experts could reach consensus on how best to respond. In September 2022, US President Biden, and others, declared that 'the pandemic is over', which was seen to reflect public sentiment in the US that the virus was receding 'even as hundreds of Americans continue to die of covid each day' – but many public health authorities disagreed and warned that the claim undermined efforts to secure funding (Diamond, 2022). Then, on 5 May 2023, the WHO announced that it would no longer classify the COVID-19 pandemic as a 'public health emergency of international concern' (Taylor and Diamond, 2023) – at a time when many people around the world were still suffering or dying from the disease, and societies continued to be affected in major ways by the cascading impacts of pandemic measures.

As many experts have argued, societies are always in a 'transition phase' or 'between pandemics', and that there is no room for complacency. One of the important insights gained from COVID-19 is that much is at stake in this politics of framing a pandemic, including its origins, magnitude, and duration, for people's lives, livelihoods, and their futures. It is crucial, then, to delve into the details of how this emergency or crisis has been defined and how this has shaped the responses.

Characteristics of a crisis

For a sociologist, the question of how one classifies and defines a phenomenon is of great consequence for how people respond to it. Epidemics and pandemics, like earthquakes, hurricanes, floods, major fires, and tsunamis may seem to be 'natural' catastrophic events, and their significance beyond doubt, but they are socially constructed and framed in ways that invite certain responses. To say that a phenomenon is constructed is not to deny its materiality. More than two decades ago, the philosopher Ian Hacking (1999), suggested that the constructivist metaphor has perhaps been overused in the social sciences and may suggest that the phenomenon that is said to be constructed has no reality and consequence for those involved. The question Hacking rightly poses is what, precisely, is being constructed? To say that crises are constructed is not to deny their manifestations or material effects but, rather, to acknowledge that they could be defined in many ways in terms of their dimensions and implications. Consequently, it is *the idea* of a crisis that is at issue; that is, how a crisis is represented in policies, public statements, media reporting, and so on.

Crises serve to expose hidden features of the social order, especially inequalities and the structures that produce and sustain them. This may happen directly by drawing attention to the different impacts of those events; for example, people's inability to shield themselves from the harms that result from armed conflict, or the prevalence of poor living conditions experienced by some communities that make them especially vulnerable to the impacts of climate change, natural catastrophe, or disease. It also occurs via the social responses that are made to manage or lessen the consequences of the events and prevent their recurrence, which advantage some groups and disadvantage others. The latter includes the failure to address issues that may have contributed to the crisis. These responses reveal a society's values and priorities, the interests involved, and the issues at stake. As the sociologist Sylvia Walby (2015) observes, crises may have major consequences, including systemic breakdown or a route to renewal, although there is no certainty of any particular outcome since uncertainty regarding the outcome of a crisis is integral to its definition (2015: 31).

As Walby argues, crises are distinguished by their *cascading effects*, as can be seen with COVID-19 and the supply chain shocks and inflationary pressures resulting from the combination of long lockdowns, closed borders, governments' stimulus measures, and the energy crisis linked to the war in Ukraine. The measures adopted by central banks to stimulate economies following control measures that disrupted supply chains and froze markets – and the recessionary impacts that followed – have had significant cascading impacts, as I explain in Chapter 4. Observation of these effects challenges the idea that crises are time-limited events. Recognizing that a pandemic crisis, or indeed any crisis, has reverberating effects brings into question the idea that the consequences of related disruption can be understood within short-to-medium time frames; for example, in evaluating the number of illnesses, hospitalizations, and deaths.

The word 'disruption' has been widely used to describe the impacts of the COVID-19 pandemic, especially in the early months of its unfolding, namely the alteration of taken-for-granted aspects of life. This disruption happens on many levels and may have lasting impacts on some groups. If one takes as a measure the level of job losses, a US assessment of COVID-19 in its first year concluded that 'No other recession in modern history has so pummeled society's most vulnerable', and that while the downturn for educated white-collar workers quickly rebounded after the peak of the pandemic-induced recession (February to April 2020), low-paid workers, who include largely Black people, women (especially mothers of young children), and young people, continued to suffer high levels of unemployment (Long et al, 2020). Women suffered more than men by job losses due to their employment in those service industries that were immediately affected by lockdowns, more inclined to perform the 'triple shift' of paid work, domestic work, and emotion-based labours of looking after children who were required to undertake home-based schooling, and more vulnerable to domestic violence during these periods.

A report of the UN published in 2021 predicted that while COVID-19 would push more people into extreme poverty, women and girls would be the hardest hit, given that many would have prioritized family obligations over paid work, which can adversely affect their incomes in their working years (Azcona et al, 2021). Subsequent evidence produced by the World Bank (in 2022) and Oxfam (in 2023) confirmed this prediction, as well as the increase in extreme poverty and massive disparities in wealth (see Chapter 4). Older citizens in aged care facilities, many with comorbidities, were especially prone to the risks of COVID-19 infection. This infection was often transmitted by workers in casual, low-paid employment with little or no training and experience in infection control who spread disease across different facilities as they sought to juggle multiple jobs to earn a decent living (see, for example, Brainard et al, 2020; Royal Commission into Aged Care Quality and Safety, 2020). Older people experienced high rates of infection and deaths in the early stages of the pandemic – an outcome of multiple failures that highlight the ageism and age-based inequalities that beset many societies.

On the other hand, the COVID-19 pandemic benefitted some individuals and groups, sometimes greatly, even if only temporarily: families reporting spending more time together during lockdowns, the move to more flexible, 'family-friendly' work arrangements or working from home, and greater reflection on personal priorities and the opportunity to slow down or begin new projects, and/or change careers or place of residence. While many businesses suffered or went broke during the COVID-19 lockdowns, and some groups became poorer, others thrived and profited, often massively, especially Big Tech and companies that utilize their products in their businesses, and those companies able to provide goods needed by consumers during the long periods of lockdown, such as groceries and homeware items (see Chapters 4 and 5).

In short, the disruptions of COVID-19 were multifaceted, with highly variable and inequitable outcomes. Modernization, globalization, and digitization have meant that nations and cultures are now highly interconnected and interdependent, with technology playing a critical role in mediating experiences and efforts to mitigate and manage the outcomes of pandemics. This interconnection has enabled both the virus and information about its risks and methods of control to spread at a rapid rate. Yet, news coverage in the richer countries mostly located in what is often described as the Global North neglected the plight and the concerns of those living in the poorer countries of the Global South, including Haiti, Yemen, and Sudan, and the Middle East and Central Africa, which are compelled to constantly contend with endemic disease, political chaos and conflicts, famine, and a daily struggle for survival, which were often more pressing than COVID-19 (Fassin and Fourcade, 2021). During the pandemic, expressions of international 'solidarity' often vied with nationalist impulses and self-interest – as revealed by the frequent mismatch between authorities' stated intentions and policies and programmes. At the national level, responses and outcomes often varied greatly as well. For example, in Australia the Federal Government was often pitted against state governments in deliberation on responses, such as lockdowns (Le Grand, 2022).

While some of the trends I describe were evident before COVID-19, pandemic conditions served to accelerate them, such as growing inequalities and the advance of AI, as I argue in Chapters 4 and 5. However, while the pandemic has shone a light on many issues, there is no certainty about how pandemic societies will evolve in the future. Recognition of the ever-present potential for change provides ordinary citizens with scope to reimagine that things could be otherwise and perhaps not just more effective in terms of tackling the disease itself but also for changing conditions that predispose to pandemics, including inequalities and human-induced climate change, as I discuss in Chapter 6.

Emotional regimes of a crisis

As I elaborate in Chapter 2, the framing of the COVID-19 pandemic as a crisis called for certain *affective* responses, or distinct ways of feeling or expressing emotions. A sense of urgency, fear, and uncertainty were conveyed in official communications and public representations of the pandemic which extensively used military analogies to emphasize the gravity of the threat. These representations included government-sponsored marketing campaigns that employed fear-laden messages, and media images of rising infection and death rates, hospitals in chaos, and rows of coffins of those who died from or whose death was contributed to by COVID-19 waiting to be buried (for example, the mass burials on Hart Island, New York City) which were widely circulated via television and the internet in early 2020. At the same time, these messages also implicitly offered hope to those who complied with advice. As socio-political phenomena, pandemics are governed by distinct emotional regimes that allow scope for the experience and expression of only certain authorized emotions. Political orders are characterized by distinct emotional regimes that offer different strategies of emotional management, and different degrees of emotional liberty and self-control as well as penalties for violating norms of emotional expression (Reddy, 2001).

At the beginning of COVID-19, authorities at the WHO and the UN sought to mobilize emotions in order to engender support for the adopted measures: to begin with, fear of the virus and the illnesses and deaths that would follow from the failure to follow certain measures, and then hope in risk management tools and technologies such as purported life-saving vaccines and self-tests that would assist citizens to manage their relationship to risk and enable them to live disease-free lives.

The public representations of COVID-19 were strongly influenced by its initial definition as a disease caused by a *novel virus*, SARS-CoV-2, with uncertain characteristics and unlike that which causes seasonal flu. This provided the context for the generation and circulation of frightening stories of potentially uncontrollable contagion. The first reported cases of COVID-19, the coronavirus disease caused by SARS-CoV-2, were in Wuhan City, in China, in December 2019. The virus is described in documents produced by the WHO as genetically related to a large family of coronaviruses, some of which cause respiratory diseases in humans such as Severe Acute Respiratory Syndrome (SARS), which was detected for the first time in 2003 (World Health Organization, 2020a). The definition of COVID-19 as viral based is not just scientifically descriptive but has distinct connotations that have impacted policy responses and public reactions.

Medical sources identify the following properties of a virus:

- They are microscopic, and much smaller than bacteria (80 nanometres versus 1,000 nanometres (1 nanometre = one thousand-millionth of a metre).
- They are responsible for some of the most devastating human diseases, including influenza, smallpox, and poliomyelitis.
- They cannot grow or reproduce outside a living cell.

- They invade living cells and use their 'chemical machinery' to survive and reproduce themselves.
- They may reproduce with errors (mutations), making treatment difficult.
- They spread from person to person either directly, for example, through breathing in airborne droplets, sexual contact, or eating food or drinking water contaminated with a virus, or indirectly via a host such as a mosquito, mouse or tick.
- They may originate with animals (zoonotic diseases), as with coronavirus disease (COVID-19 caused by SARS-CoV-2). (Marcovitch, 2017: 708; Haider et al, 2020; Stöppler, 2021)

During the 20th and early 21st centuries, this medical conception of a virus has informed popular cultural representations and community understandings of risk and danger. Films such as *Contagion* (2011), *Plague* (1979) and *Infected* (2004), for example, depict the rapid spread of invisible viruses (contagion) that are the source of fear and panic in the population. With the rise of the internet, the term virus began to be used to convey the infective potential of foreign agents in computer programs that may rapidly mutate and cause program malfunction and other harms. The term 'going viral' is used to convey how an image, story, or video may spread rapidly and widely via the internet and suggests a loss of control over initial messages.

Fear of this form of viral contagion was evident from the outset of COVID-19, with concerns raised about the threats to public health posed by the 'infodemic' of misinformation and rumours rapidly spreading via social media (Zarocostas, 2020a, 2020b) (see Chapter 2). Viruses are seen as inherently dangerous and originating with sources that *need* to be controlled. This can mean and, during COVID-19, often did mean a focus on certain groups, entities, locales, or even entire populations or countries that are perceived as potential sources of infection. This is common with infectious disease outbreaks, as had been seen earlier with HIV/AIDS, Ebola, and SARS (for example, Dowell et al, 1991; Buus and Olsson, 2006; Roy et al, 2019).

Attributing blame and responsibility

During COVID-19, efforts to ascribe responsibility and blame were seen in the extensive debate among scientists and in the news media about the origins of the coronavirus SARS-CoV-2, with some commentators pointing to the Wuhan wet markets and others suggesting it had 'escaped' from a lab in that city (the Wuhan Institute of Virology) which has carried out research on coronaviruses which are endemic in the region where it is located – the so-called 'lab leak' hypothesis (Maxmen and Mallapaty,

2021). It has been suggested that the virus may have been a product of so-called 'gain of function' research which created a transgenic virus (Warmbrod et al, 2021). ('Gain of function', it should be noted, may also occur 'naturally'.) Another, more recent theory is that COVID-19 originated with raccoon dogs at Wuhan markets. In March 2023 it was reported that an analysis of gene sequences by an international team had found 'Covid-positive samples rich in raccoon dog DNA' (Sample, 2023). According to the WHO and many national health authorities, this search for the origins of the virus will inform future public health responses. Did the virus have a natural origin, being transmitted from an animal to humans (a 'zoonotic transfer')? Was it unintentionally released from a lab, perhaps during a scientific experiment? Or was the virus a result of a botched Chinese bioweapons programme – or brought to Wuhan by the US military as a bioweapon (Knight, 2021)?

In February 2023, debate on the 'lab leak' hypothesis was rekindled when it was reported that an updated version of the US intelligence community's report to President Biden indicated that 'the Energy Department has shifted from a neutral stance on the virus' origin to the one favoring, with "low confidence", a lab leak' (Achenbach, 2023). (The Energy Department runs major national laboratories and invests billions on scientific research.) It is unclear why the Energy Department changed its view, since the updated intelligence report remains classified. This report noted that the Wuhan Institute of Virology was the focus of the lab leak conjectures since it is a major research centre undertaking research on coronaviruses, for which they collect wild bats, the source of coronaviruses, and that someone involved in the research may have inadvertently introduced the virus into the population (Achenbach, 2023). It has been suggested that potential 'gain of function' experimentation, or some other type of viral manipulation, could have led to the creation of SARS-CoV-2. The construction of the COVID-19 pandemic has been characterized by many such origin stories, which have been critical in shaping public responses.

From the outset of the pandemic, these origin stories were highly politicized, with Chinese authorities, for example, arguing that the virus in fact originated in the US, having leaked from the military base in Maryland (BBC News, 2021a). Scientific and public debate on these questions helped fuel doubt and suspicion, fuelling conspiracy theories and rumours, which rapidly spread through social media (Knight, 2021). Viruses instil fear and engender efforts to attribute responsibility and blame. Such efforts can quickly slide into the latter, with certain groups being subject to heightened suspicion, surveillance, and control. During COVID-19, blame was implied in news media reports and policy discussions about disease outbreaks in certain areas (generally lower socio-economic) and among certain communities (typically minority ethnic).

Risk governance and its failures

Official responses to COVID-19 and other recent epidemics and pandemics have been consistent with an approach that scholars refer to as 'risk governance'; that is, 'the "translation" of the substance and core principles of governance to the context of risk and risk-related decision-making' (Guggingham et al, 1998, cited by Renn, 2008a: 8). As I discuss in Chapter 2, risk governance has underpinned policy making in many domains including public health and medicine. The limitations of this approach became especially evident towards the end of 2021 with the emergence of the highly contagious Omicron variant and the failure of existing measures involving centralized systems of monitoring ('track-and-trace', use of QR (quick response) codes) and testing using the 'gold-standard' PCR (polymerase chain reaction) tests. From around this time, many governments adopted a different strategy, with an emphasis on self-responsibility for the management of risk, a central element of which is mass self-testing using rapid antigen tests (Petersen and Pienaar, 2024).

It is now well understood that COVID-19 is caused by a virus surrounded with many *uncertainties*, including its disease processes, origins, infectiveness, transmissibility and mechanisms of transmission, its longer-term health effects, the differential vulnerability of different groups to the disease, and effectiveness of treatments (Koffman et al, 2020). The evolution of variants and new strains of the virus have complicated control efforts. Uncertainties also surrounded people's responses to interventions, many of which restricted rights of movement and association, and mandated certain practices such as vaccinations and use of facemasks. The uncertainties of COVID-19 provided scope for different groups to impose their own interpretations of the pandemic and required responses – which rapidly circulated online – which underlined the limitations of risk management and mitigation measures.

As Leach et al (2022) argue, risk management approaches used in public health, such as those guiding pandemic preparedness and planning, assume a predictable, controllable future, which fails to address uncertainty as well as ignorance (Leach et al, 2022). Embracing incertitude in disease preparedness, the authors contend, demands that social, political, and cultural dimensions are made central in policies. This includes recognition that diseases (like COVID-19) evolve in unpredictable ways and have unknown, longer-term implications, and that the responses to both the disease itself and to interventions cannot be easily foreseen, being shaped by factors such as the priorities and beliefs of those subject to interventions, and historical trust or distrust in authorities to act in the best interests of citizens (Leach et al, 2022).

Modelling

Critical to the task of mitigating and managing risk during COVID-19 was the use of modelling involving a range of techniques and technologies which promise to render the pandemic intelligible and potentially controllable. Following the onset of the pandemic, the language of modelling quickly became part of policy and popular discourse on the pandemic. News articles and television coverage made numerous references to modellers' predictions, often explained by epidemiologists and other experts assisted by visual displays, including the use of graphs, figures, charts, tables, diagrams, and other techniques – what Bruno Latour and Steve Woolgar call 'inscription devices' (1986: 51). As Latour and Woolgar have argued, these inscription devices are a means not so much of transferring information but of *creating order out of disorder* (1986: 245). Models, then, can be considered technologies of order making. They are a product of modernization and characteristic of modern states and the 'expanding wave of information gathering practices' that have been their defining characteristic (Bowker and Star, 2000: 117). As I explain in Chapter 3, epidemiological modelling has a long history, but the approaches that are now commonly used evolved in the early 20th century.

One of the most used models in public health today – the Susceptible-Infectious-Recovered (SIR) model – was introduced in 1927, less than a decade after the 1918 influenza (Tolles and Luong, 2020). The SIR model is one of the simplest so-called compartmental models used in the modelling of infectious diseases, whereby the population is assigned to and progresses between 'compartments' with labels such as 'S' (the number of susceptible individuals; that is, those exposed to the pathogen), 'I' (the number of infectious individuals, or those colonized by the pathogen) and 'R' (the number of recovered (previously infected and consequently immune) or deceased individuals), with the order of the labels indicating the flows between the compartments. The model is used to predict the spread of disease, the numbers of people infected, the duration of an epidemic or pandemic, and the impacts of different public health interventions such as vaccination rates and restrictive measures on the outcome of an epidemic (see, for example, Kermack and McKendrick, 2017; orig. 1927) (see Chapter 3).

Like wars, the disruptions of epidemics and pandemics often serve as a catalyst for social and technological change, including of the technologies and techniques of modelling – as has been evident during COVID-19. Regardless of the model adopted, all involve 'trade-offs' – the assumed consequences of pursuing particular strategies in terms of their costs and benefits, with those judged by modellers and policy makers as most likely to benefit 'the public's health' (as measured by reduced infections and/or deaths) generally favoured over other options (including doing nothing).

This utilitarian approach tends to disregard the variable impacts of decisions on different groups and communities.

During COVID-19, public health experts provided daily updates on infection rates, predicted patterns of transmission associated with different measures, and the economic and social impacts of lockdowns (see Chapter 3). The use of visual material such as infographics to depict trends and projections based on different scenarios according to levels of vaccine coverage, the wearing of facemasks (when it became clear that the virus was conveyed via aerosol rather than on surfaces, as many experts originally thought to be the case), the spread of variants, and other factors, conveyed both the issues at stake in making different decisions and belief in the calculability and controllability of risk. Epidemiologists, immunologists, and chief medical officers achieved a high public profile through their appearances in rolling media coverage and news releases – their constant presence both confirming the severity of the threat posed by the virus and offering reassurance that citizens were in safe hands.

Many of these experts became celebrities and praised for their wisdom, calmness, and measured advice. They, along with vaccine developers, such as Professor Sarah Gilbert at Oxford University, and hospital staff who dealt with the mounting numbers of illnesses, were treated as 'heroes' or 'heroines' who saved many lives. Such depictions are performative in the sense that they helped make real the future scenarios portrayed by models and the dangers that lay ahead should advice not be followed. They were critical in producing responsible citizens, who should remain constantly alert to the prospect of infection and the need to play their part in protecting themselves and others.

Within a relatively short period of time, citizens were called upon to assimilate new classificatory schemes and language into their world-view to assist them to understand and manage their own relationship to risk. While many hopes have been attached to these models and the credibility of the science upon which they rely, their implicit classificatory schemes have not been stable; rather, they shifted in line with changing scientific knowledge of the disease and its transmissibility, its paths of transmission, reproduction rates (the 'R-value' or 'R number'), and other factors. One consequence of this was that people's hopes and expectations for the future changed as the pandemic evolved.

Uncertainty is an intrinsic feature of virtually all mathematical models, as modellers themselves acknowledge. Yet, the digital technologies upon which models rely to collect, aggregate, categorize, and analyse large pools of data ('big data') in real time promise to provide a definitive picture of different scenarios and viable options. Real-time data analytics made possible the development of a global surveillance infrastructure that would have been unthinkable to governments and public health authorities confronting pandemics in the past. The WHO and its Collaborating Centres, the Institute for Health Metrics and Evaluation (IHME), John Hopkins University, and other universities, organizations, and individual researchers produced online 'dashboards' or 'resource centres' comprising up-to-date visual representations of trends and projections that were readily available to researchers, public health authorities, and the wider community.

This data was used for various purposes, including to assess the risk and resilience of different countries and regions, and chart daily infections and testing, mask use and social distancing, hospital resource use, and daily deaths (for example, IHME, 2021). The WHO, for example, produced a 'Coronavirus (COVID-19) Dashboard' that offered continual updates on global trends on cases, deaths, and vaccine doses, broken down by WHO region, with snapshots of different countries, territories, and areas (World Health Organization, 2021c). This avalanche of digitally mediated data and visual depictions of COVID-19 has transformed not only people's understanding of pandemics but how people understand their societies.

Summary of the remaining chapters

In Chapter 2, I discuss sociological and other writings on crises, highlighting their relevance for understanding the official framing of and responses to the COVID-19 pandemic and other pandemics. This work uncovers some typical features of such events including their disruptive and cascading impacts, the significance of emotional responses in shaping actions, and the extension of authorities' powers and suspension of citizens' rights associated with the 'state of exception'. I draw on this work to explore the dynamics of framing the COVID-19 pandemic, beginning with the UN's call for concerted, coordinated action and initial subsequent responses. The chapter identifies the kinds of expertise enrolled to help manage related disruptions that threatened economies and social stability. The 'COVID-19 crisis' soon exposed tensions and contradictions in the social order and the limitations and failures of risk governance which relies on science- and technology-based approaches. These failures were manifest in resistance to measures and the so-called infodemic of alleged misinformation and rumours. While similar forms of resistance have been apparent with earlier pandemics, digital media have enabled diverse information to travel further and faster than previously. Countering and controlling the flow of such information was an integral part of efforts to manage the pandemic. I conclude by noting that, while the full implications of the COVID-19 crisis will not be known for many years, by May 2023, when the WHO declared the end of the pandemic, its effects continued to reverberate, with disruptions to supply chains and labour and high inflation in many countries.

In Chapter 3, I explore the critical role played by modelling during the COVID-19 pandemic and the implications of policy makers' reliance on modellers' predictions. The WHO's declaration of an emergency in 2020 brought to public prominence the arcane language of epidemiology and public health. The pandemic has arguably altered community perceptions of pandemics and infectious disease and of public health experts. In the chapter I discuss how modelling shapes both understandings of and responses to a pandemic and conceptions of science and society. As I argue, models do not just predict potential futures but *shape* futures in ways that are rarely acknowledged in debates about their utility. The chapter examines the different kinds of models used in public health, their biopolitical and surveillance implications, and the history of their use in earlier epidemics and pandemics. Models can be conceived as 'crisis technologies' that have many shortcomings in practice, and their applications well illustrate science-as-politics. I also consider shifts in the practices of science and technology manifest in 'real-time' modelling during COVID-19, and the various initiatives used to collect, share, and utilize data. While some scientists have raised concerns about reliance on modelling for pandemic decision making, notably when used to justify lockdowns, as occurred in 2020, these concerns were mostly ignored or set aside in the enthusiasm to embrace new technologies based on the utilitarian calculus of the presumed community benefits that would be derived from measures.

In Chapter 4, I examine the inequalities made evident or reinforced by 'the COVID-19 crisis', drawing on research published in the three years after its onset. I revisit the UN's initial plea to the rich G20 countries to make a coordinated response to the pandemic, and evaluate subsequent responses. As I note, the UN and the WHO outlined the massive economic and social disruptions of the pandemic, making a plea for global solidarity led by the G20, which met with a limited response. While many governments implemented stimulus measures, in line with the UN's call, support packages were selective in their application, and the measures have had inequitable impacts. Three years after the declaration of the emergency, research showed that stimulatory measures had been excessive and fuelled inflation, which affected businesses and citizens' standard of living. They also disproportionately negatively affected poorer countries, which contributed to the creation of a 'two-speed global economy'. The chapter discusses the factors that contributed to the failures of the G20, manifest in unequal access to vaccines and so-called vaccine nationalism – a key issue in the second year of the pandemic. The chapter also examines critical evaluations of pandemic preparedness and response undertaken by a WHO-commissioned panel, and other evidence on growing wealth disparities linked to COVID-19-related measures.

In Chapter 5, I outline the dimensions and socio-political implications of the technological changes wrought by pandemic crisis conditions. As I explain, COVID-19 served to accelerate the collection and sharing of data – and consequently the rapid advance of AI – which assisted pandemic surveillance and encouraged massive investment in digital technologies. However, while the intensification of data sharing during COVID-19 proved invaluable for researchers and for ordinary citizens compelled to work and undertake education at home during lockdowns, it also proved favourable for various state and non-state actors to undertake cybercrime and other nefarious activities. The crisis provided the catalyst for a technological 'arms race' among Big Tech to corner the AI market, reflecting a 'winner-takes-all' conception of technological development that potentially has far-reaching impacts on work, education, and decisions about where to live. The chapter outlines some major trends induced or accelerated by the crisis, such as the rapid integration of videos into people's lives and the launch of various new AI-assisted innovations. Two tangible indicators of this change are the massive investment in data infrastructure and increased focus on AI in national research priorities. The rapid, real-time sequencing of the SARS-CoV-2 virus during COVID-19, I argue, has heightened optimism for the utility of AI in future pandemic responses.

In Chapter 6, I consider what might be learnt from 'the COVID-19 crisis' for societies confronting future pandemics and particularly public health. I ask, could pandemics be framed differently to how COVID-19 was framed to avoid similar damaging effects? I begin by examining the issues at stake in the framing of an event as a crisis in the light of emerging evidence on COVID-19. I then outline the kinds of changes I believe are needed to better prepare societies to meet the challenges posed by future pandemics. This includes reforming the global governance of public health, which was found wanting during COVID-19. I argue for a new kind of public health: one that is less science and technology focused than current public health, and that pays greater acknowledgement to uncertainty, the significance of emotions in responses, the potentially inequitable impacts of measures, and utilizes the media to broaden the debate on the critical issues. Finally, I consider the prospects for social renewal provided by conditions of converging crises, including climate change which may precipitate infectious disease outbreaks in the future. Sociology and other social sciences, I suggest, can contribute much to the project of renewal which will assist to meet the challenges posed by recurrent pandemics and avoid the kind of harmful outcomes of responses seen during the COVID-19 pandemic.

2

The Politics of Framing a Pandemic Crisis

As argued in Chapter 1, pandemics are not just biological phenomena but also socio-political events. As such, the question of how they are defined is critical for the character of the responses. In this chapter, I discuss the dimensions and implications of the politics of the framing the COVID-19 pandemic 'crisis', focusing on the contending claims of supranational and national authorities and other actors regarding measures that restricted citizens' freedoms to varying degrees. The pandemic served to expose the tensions and contradictions in the social and political order in many countries, especially the difficulties that authorities face in managing competing economic and social objectives during a period of a defined crisis. It also highlighted the limitations of the prevailing 'risk governance' approach, which is premised on the idea that the future can be predicted and controlled (Renn, 2008a, 2008b). Risk governance in its idealized portrayal entails centrally coordinated systems of surveillance, pre-detection (control before the event), and prevention, which denies uncertainty (especially regarding human–technology interactions), ambiguity, and failure (for example, in technologies and systems) – all of which have been evident during the COVID-19 pandemic.

I begin by examining some major sociological writings on crisis which prove invaluable in helping to make sense of social responses to COVID-19 and other pandemics. The politics of framing a crisis, I propose, involves various dimensions and dynamics which have been apparent during the COVID-19 pandemic: 'the state of exception' and suspension of established rights, appeals to 'health' and fear of disease, experiences of disruption, the operation of emotional regimes, the questioning of identity and taken-for-granted assumptions, and the uncovering of inherent systemic tensions. I then illustrate my argument with reference to a critical juncture in the unfolding of the COVID-19 pandemic when the WHO and the UN endeavoured to frame the event as a crisis, calling for concerted action. From the outset of the pandemic, different vested interests sought to shape the public representations of the crisis – its character, magnitude, and consequences – and hence the agenda for policy and action; for example, efforts to reopen economies and/or exploit crisis conditions to advance certain objectives. In conclusion, the chapter will consider the implications of this analysis for understanding responses to pandemics of a similar or larger magnitude to the COVID-19 pandemic in the future.

Analyses of crises in sociology

Since the beginning of their discipline, sociologists have been interested in the dynamics of crises and their impacts on societies. This work can be used to help shed light on the responses to pandemic-related disruption of the kind experienced with COVID-19 and, potentially, future pandemics. Writers such as Emile Durkheim, Karl Marx, and Max Weber and, more recently, Jurgen Habermas, Stuart Hall, Wolfgang Streeck, and Sylvia Walby have made significant contributions to our understanding of crises – especially economic, political, and legitimation crises. Not all these scholars directly address the question of what exactly constitutes a crisis, let alone a pandemic crisis, but all share a concern with their predisposing conditions and the economic, social, and individual consequences.

Walby offers a useful starting point for analysing a crisis, with reference to the financial crisis of 2007–08 and its cascading effects on the economy, politics, and society. As she observes:

> A crisis is defined … as an event that has the potential to cause a large detrimental change to the social system and in which there is lack of proportionality between cause and consequence. Crises are both 'real', in the sense of actual changes in social processes, and socially constructed, in the sense that different interpretations of the crisis have implications for its outcome. The interpretation of a crisis may under- or over-state its magnitude and impact, as well as attribute blame as to its cause. (2015: 14)

While one may take issue with Walby's distinction between 'real' and 'constructed' in that it suggests the existence of an unmediated reality, her definition is valuable in acknowledging the significance of how crises are 'framed' for the character of the responses, which then affects how they unfold. It also implicitly acknowledges the potential for different actors to shape the discourse on the causes, magnitude, and impacts of a crisis – and hence the character of subsequent responses. As Walby observes, while the literature uncovers both realist and constructivist approaches to crises, all 'share the assumption that there is a relationship between power and knowledge; in particular, that power shapes knowledge' (2015: 15).

Walby notes that in times of crisis 'the range of meanings under consideration opens up, only to narrow again as social relations take new forms of institutionalization' (2015: 34). Lurking behind the concept of crisis, she argues, is system breakdown, and consequently disaster or catastrophe. Stories of crisis involve accounts of loss and may impel efforts to re-establish 'normality'. They also carry the potential for large-scale or systemic change. As Walby explains, a crisis 'might provoke a break in the existing trajectory of development and the creation of a new, path-dependent trajectory of development' (2015: 34). She suggests this might result in a different model of capitalism or gender regime.

The purported significance and impacts of crises will vary according to the power of competing interests/claims-makers to shape knowledge and establish the agenda for policy and action. There is, then, no 'blueprint' for how crises evolve, which will be shaped by competing interests that seek to impose their 'definition of the situation'. While authorities may 'frame' a crisis in a way that serves to mobilize community support for measures to alleviate its effects or to allay related fears and panic – evidence of which was seen during COVID-19, as will be examined in detail later – there is always the potential for authorities to lose control of representations, which may have unintended consequences. Indeed, as I will argue, this has been apparent with COVID-19 when different actors sought to shape views and outcomes in ways that called into question the legitimacy of authorities' rule.

COVID-19 and experiences of crisis

For people around the world, the sudden onset of the COVID-19 pandemic in the early months of 2020, involving national lockdowns in many countries – which extended into 2021 for some jurisdictions such as Victoria, Australia, and December 2022 for China – was unexpected, and for many people highly disruptive and unsettling. Widely portrayed in news and other media as an 'unprecedented crisis', the measures inflicted a kind of collective shock, with whole economies suddenly shut down with many people losing their jobs or businesses and facing uncertain futures. Students' schooling was also disrupted, often for extensive periods, as they and their teachers and parents were compelled to suddenly come to grips with the demands of online teaching and learning and, in many cases, the limitations of home environments and digital communications, among other challenges. The dimensions of this disruption and related changes have now been widely documented – although details of events in economically poorer parts of the world are less well known to those in the relatively resource-rich parts of the world (apart from the harrowing, fear-inducing news media images in the early phases of COVID-19).

People in precarious employment, such as those in the gig economy, visual and performing artists, and hospitality workers, were among those greatly impacted as they suddenly lost their customers or audiences, and consequently their livelihoods and, for many, their sense of purpose. Extended lockdowns have had various other outcomes in the short-to-medium term, including increased rates of domestic violence, self-reported loneliness and diagnosed mental illnesses, delayed diagnoses and treatments, and increased reliance on digital media for communications. The longer-term implications of pandemic policies on individuals and societies, let alone of the disease itself, are immeasurable, and the full extent will not be known for many years.

While for most people around the world the pandemic was unexpected, before COVID-19 supranational bodies such as the WHO, the World Economic Forum and the World Bank warned of the risks posed by future *economic and social disruptions* of pandemics, noting that countries lacked 'pandemic preparedness' (for example, Jonas, 2013; World Economic Forum, 2019). In 2013 the World Bank predicted that there would be 'possible demand and supply shocks' associated with what was then envisaged to be a likely flu pandemic, and that 'most of the impact [from a flu pandemic] would be due to people's reactions and avoidance behaviours', which would be 'aggravated by likely confusion triggered by incomplete or inaccurate information and other inadequacies in individual subjective risk assessments' (Jonas, 2013: 3). These reports convey concerns not just about the economic and social disruptions of pandemics but about authorities' ability to ultimately manage or mitigate the risks through 'preparedness'.

State of exception

As the Canadian author-activist-filmmaker, Naomi Klein, writes, in *The Shock Doctrine* (2008), crisis events, manufactured or 'natural' (for example, major cyclones, floods), provide the precondition for societies forfeiting things which they would otherwise fiercely protect, including human rights. Klein draws on multiple examples of how crises enabled the instalment of neoliberal policies in many countries around the world, with proponents using a combination of 'shock therapy' and violence to achieve their goals. In March 2020, at the beginning of the COVID-19 pandemic, Klein released a video outlining her views on how 'the still-unfolding Covid-19 crisis is already remaking our sense of the possible' and commented that 'The Trump administration and other governments around the world are busily exploiting the crisis to push for no-strings-attached corporate bailouts and regulatory rollbacks' (Klein, 2020). Authorities in many countries exploited the widespread sense of chaos and fear to implement programmes and measures that would otherwise have been difficult and perhaps impossible to implement. The declaration of the COVID-19 pandemic provided the grounds for the introduction of emergency powers and measures such as lockdowns and border closures that would be unthinkable in other circumstances. The 'state of exception' that followed the declaration has clear parallels with the emergency powers introduced by the Bush Administration following the attacks of 9/11.

Many businesses and individual entrepreneurs also exploited the crisis to generate profit. Companies made billions of dollars from the chaos and gyrations of the stock market by, for example, purchasing and selling insurance contracts known as credit default swaps (Patterson, 2023). In business circles, it is well known that extreme unexpected events – or so-called Black Swan events – such as stock market crashes and financial crises are not just to be expected and planned for but create an opportunity to generate profit. (For insight into the dynamics of the stock market in the age of pandemics, see Patterson, 2023.) Many companies, especially technology companies, massively increased their profits during the COVID-19 pandemic. The crisis conditions served to widen disparities in wealth, as I discuss in Chapter 4.

Yet, while such measures may seem exceptional or without precedent, as Italian philosopher Giorgio Agamben (2005) observes, they are characteristic of a form of rule evident in many countries today. Writing in 2021, a year into the pandemic and referencing events in his own country, Agamben noted that the 'disproportionate response' to the pandemic involving the suspension of citizens' rights 'for public health and security reasons' has become a 'standard paradigm of governance' (2021: 13–14). At the beginning of his book, he argues:

> Almost a year now into the emergency, we should consider the events we have witnessed within a broader historical perspective. If the powers that rule the world have decided to use this pandemic – and it's irrelevant whether it is real or simulated – as a pretext for transforming top to bottom the paradigms of their governance, this means that those models were in progressive, unavoidable decline, and therefore in those powers' eyes no longer fit for purpose. (2021: 5–6)

Agamben comments that a similar state of crisis pertained during the third century of the Roman Empire that 'launched a series of radical reforms of its administrative, military, and economic structures' that ultimately led to the Byzantine autocracy. The defining feature of the pandemic-induced transformation, he argued, is the suspension of constitutional guarantees. Agamben likens this state of exception – characterized by the 'introduction of a sanitation terror and a religion of health' – to that declared by Adolf

Hitler in 1933, which lasted for 12 years and that 'was necessary to deploy an explicitly totalitarian ideological apparatus' (2021: 7–8).

History of appeals to 'health' and fear of disease

Agamben's comparison of the COVID-19 crisis with the period of Nazi Germany may seem extreme. However, religious-like appeals to 'health' and fear of disease have long been used to justify legislation limiting personal freedoms and to persecute certain groups. Paternalistic programmes involving 'medical police' were introduced in late 18th- and early 19th-century Europe, to cover all aspects of life, including municipal planning, housing, sanitation, water supply, sewerage disposal, child hygiene, and school planning (Dubos, 1996: 212–213). As Dubos argues, 'in the past, problems of health have provided compelling reasons for restricting freedom' (1996: 236).

During the 19th century, knowledge that diseases disproportionately affected economically poorer groups soon convinced the wider public as well as industrialists and politicians to undertake concerted social action against the threat of disease. As Dubos explains, it 'created a political climate in which the preservation of the health of the masses came to be regarded as the responsibility of the community as a whole' (1996: 236–237). Fear of infectious disease, and especially cholera in the mid-to-late 19th century, stimulated the creation of national health authorities and centralized laboratories, giving 'ever-increasing power to health departments for the regulation of community life' (1996: 237).

In the 20th century, the growing influence of genetics provided the foundation for state eugenic programmes in many countries, reaching its apotheosis in Nazi Germany. In his book, *Life Unworthy of Life*, James Glass describes German officials' obsession with rebuilding a genetically fit race in the aftermath of the losses of the First World War (1997: 31). As Glass explains, many officials, scientists, and administrators feared the genetic effects of chronic disease, mental disorders, and sexual decadence, and the new science of genetics gained adherents across many sectors of society. It was commonly believed that the biological integrity of the race was under siege and that 'correct' action was required.

As Glass explains, by the 1930s 'the eugenics that emphasized collective health gave way to a eugenics focusing on the dangers of racial and blood pollution' (1997: 32). Many municipalities in Germany administered programmes of 'racial hygiene' with the goal of regaining the health of the people. Despite some pockets of resistance to this 'cultural obsession with eugenics', biological science became absorbed into politics, affecting both left and right, and led to the implementation of policies that included 'the violent elimination of selected groups as a public health policy' (1997: 35).

As these histories show, the use of metaphors, particularly those of medicine and public health, and the language of 'hygiene' and the fear of 'contagion' have been critical in mobilizing support for discriminatory, and sometimes genocidal, regimes. During the Nazi period, certain groups, especially Jewish people, homosexuals, and the mentally ill, were posited as the source of infectious diseases and dangerous – and not understood as human beings but as 'life unworthy of life' – which should be 'cleansed' from the body politic with the assistance of the biological and medical sciences (Glass, 1997).

Disruption

One of the key themes in sociological and other writings on crisis is the accompanying social and personal disruption. As I noted in Chapter 1, the use of 'disruption' has appeared frequently in COVID-19 narratives and in the stories that people tell in recounting the impacts of the pandemic on their lives. While massive disruptions are common in history and impact on the lives of many today, for example those who are in the midst of a war, famine, flood, or economic or political crisis, COVID-19 was an event that was widely experienced as exceptional; that is, out of the ordinary. For individuals, disruption is often accompanied by the experience of loss and despair, and a search for meaning and perhaps for someone or something to blame. The disorientations and confusion that follow from this disruption provide the occasion for authorities to implement measures that would otherwise be unpalatable and likely to be resisted by citizens. It is important to acknowledge that the meanings and experiences of disruption are historically and culturally variable.

In writing about stories of disruption long before COVID-19 (and the 9/11 attacks), Gay Becker (1997) notes that within the Western tradition notions of disruption have changed in line with shifting evaluations of order and chaos. She argues that while, within the predominant binary logic, order is positively valued and chaos is negatively denoted, there has been a shift with the advent of chaos theory. This theory suggests that 'deep structures of order are hidden within the unpredictability of chaotic systems, that chaotic systems exhibit orderly disorder rather than true randomness'. The advent of chaos theory, Becker suggests, is part of a major paradigm shift, away from the Newtonian paradigm, which emphasizes predictability and linearity, to a paradigm of 'chaotics', which 'celebrates unpredictability and nonlinearity and sees both as sources of new information' (Becker, 1997: 6). As Becker notes, it takes time for new paradigms to infiltrate daily life, especially given the pervasiveness of the old paradigm. Yet, as she argues, change is afoot and, as people begin to integrate scientific models into their understanding of their everyday world, the chaos paradigm will increasingly assume a place alongside other explanations of why things go awry.

In Becker's study of disruption in various areas of social life, including marriage breakdown, illness, and loss of employment, people were found to 'organize stories of disruption into linear accounts of chaos that gradually turns to order' (Becker, 1997: 6). Her observations, it should be emphasized, pertain to the US, where a broad section of the population hold the hope and expectation that their lives will unfold in a reasonably predictable, orderly fashion. As Becker acknowledges, in that country, the ideology of individualism focuses on continuity of the self rather than cultural continuity, and the individual, not society, is responsible for creating continuity and permanence (Becker, 1997: 99–100).

For those who live in poorer countries or regions, whose lives are less governed by individualism (or at least the extreme manifestation found in the US) and/or who come from culturally, linguistically, and sexually diverse communities, disruption may be a common and perhaps constant feature of their lives that has been merely *exacerbated* by the pandemic. For many people, chaos, unpredictability, and disruption are likely to be experienced as inherent, inescapable aspects of everyday life. Pandemic conditions, including harsh lockdowns, and related loss of employment or business, simply made matters much worse, in some cases leading to despair, mental illness and suicide or, in some cases, bitterness and resistance to measures. For others, pandemic conditions led people to re-evaluate their lives and to search for the new, expressing a hopefulness that is common in times of despair. I have more to say on hope and re-evaluation later in this chapter and in the next chapter.

Emotional dimensions

Regardless of how they originate, crises unsettle social relations and evoke strong emotions, especially fear and hope – both of which were evident to different degrees throughout COVID-19. Indeed, 'crises' are events that draw and rely on certain collective emotional responses. Emile Durkheim was perhaps the first sociologist to recognize this *emotional* dimension of crises in *The Elementary Forms of the Religious Life*, writing:

> There are periods in history when, under the influence of some great collective shock, social interactions have become much more frequent and active. Men look for each other and assemble together more than ever. That general effervescence results which is characteristic of revolutionary or creative epochs. Now this greater activity results in a general stimulation of individual forces. Men see more and differently now than in normal times. Changes are not merely of shades and degrees; men become different. The passions moving them are of such an intensity that they cannot be satisfied except by violent and

unrestrained actions, actions of superhuman heroism or of bloody barbarism. (Durkheim, 1915: 210–211)

Norberg has commented, 'Minor imperfections in vast organizational entities are not very good at claiming our attention whereas a full-scale crisis warrants emotional mobilization and demands immediate interventions' (2011: 131–132). A crisis can provide the impetus to addressing matters previously neglected and to tackling fundamental questions (Klein, 2008; see also Arendt, 1977). It may bring into sharp relief previously taken-for-granted aspects of the world and the tensions and inherent contradictions in the established social and political order. As Norberg observes, crises represent periods of both danger and loss – a loss of what people have in common, or indeed the very notion of the common, and can undermine the ground for communication (2011: 136–137).

Questioning of identity

Arendt argued that a crisis may lead to a questioning of identity and can provide the catalyst for activism and the striving for the 'utterly different', including totalitarianism (1951: 331–332). Arendt was writing in the wake of the atrocities of the Second World War and the revelations of the Nuremberg Trials (1945–49), which exposed the extremes to which humans will go when conforming to totalitarian 'group think'. Moments of collective trauma provide the occasion for radical social and economic experiments, with fear and disorder providing the impulse for radical change. This impulse for radical change was also evident during COVID-19, both at personal and societal levels – in the former in changes in employment (the 'Great Resignation') and place of residence ('sea/tree changes') for some people; in the latter, the implementation of new work arrangements and digital technologies and systems, among other innovations.

For individuals, the changes wrought by a natural or systemic crisis may lead one to question many taken-for-granted assumptions about the unfolding of one's life. As Becker (1997) explains, disruptive events affect one's sense of continuity and normalcy. A common response is to seek to create a sense of order out of disorder. This order-making may vary greatly according to the nature of the disruption, the socio-cultural and material resources available to individuals and groups to manage the disruption, and the time and place. A major life-changing event, whether divorce, a major illness, the loss of a child, a loss of employment, or the dislocations resulting from a natural or human-induced crisis, such as floods, fires, nuclear catastrophe, or group conflict, has bodily effects that shape how people see their world. As Becker explains, the body is 'the medium through which people experience their cultural world, and bodily experience can reflect the culture in which it occurs' (1997: 13).

For those living with a chronic, perhaps life-limiting illness/es, these bodily effects are perhaps most immediate and profound, and may lead to a major reassessment of one's life and prospects for the future. Individuals who live with chronic illnesses or conditions may build identities formed from a shared experience of pain and suffering and hope for improved lives in the future through access to new treatments or better-quality care and support. During the COVID-19 pandemic, most people experienced some degree of disruption in their personal lives, resulting directly from the disease (sometimes ongoing, as with Long COVID) (see Chapter 6), from loss of employment or their business, and/or from restrictions on freedom of movement and sociality. As with other major disruptive events, many people questioned who they are and perhaps have come to see themselves anew (Becker, 1997: 101).

This self-questioning may help explain why many people changed jobs, houses, and places of residence, even if temporarily (see Chapter 4). Of course, it is difficult to draw firm conclusions from shifts in patterns of employment and residence during the disruptions of COVID-19, since many factors were likely to have been at play in any set of changes, including fear of contagion in confined urban spaces. But there is much evidence to suggest that many people experienced a changed sense of self associated with feeling that their lives were out of control and a pervasive sense of uncertainty related to COVID-19 itself and their future resulting from their lives being upturned.

The disruptions of the COVID-19 'crisis' have shaped affective life in profound ways. Expressions of danger were common among those who suddenly experienced the loss of livelihood and income. The ground for communication was also disrupted, with citizens having to rely heavily on digital media instead of in-person meetings, and on multiple, often contending sources of news on the pandemic, its progress, and control measures. As noted in Chapter 1, COVID-19 was the first pandemic to occur in the digital age and this has been critical for how it has been experienced, including impacts on work, social interactions outside work, the flow of information about the pandemic, research on the pandemic itself, and the adoption of infection control measures.

Making manifest systemic crises

Yet, while those in the middle of crises may experience them as novel and unfamiliar or, as COVID-19 has been described, 'unprecedented', a claim which, I noted in Chapter 1 can be disputed, viewed historically they may also be viewed as an integral, inevitable aspect of the dynamics of social systems

34

that have been brought into sharp relief by a certain confluence of events. Some sociologists argue that crisis is a 'normal condition' of democratic capitalism, given the inherent disequilibrium and instability created by two conflicting regimes of market allocation: the 'free play of market forces', on the one hand, versus attention to social need or entitlement, legitimized by democratic politics (Streeck, 2016: 74–77). The COVID-19 pandemic arguably reinforced existing crisis tendencies in capitalist societies and made them more transparent through disrupting the fine balance between these inherently unstable regimes, calling into question the legitimacy of rule.

Writing in the 1970s, Jürgen Habermas argued that the 'crisis tendencies' that beset advanced capitalism create ongoing legitimation problems for the state (a 'legitimation crisis'). These problems, he suggests, cannot be resolved through market mechanisms alone since mass loyalty needs to be 'simultaneously secured within the framework of a formal democracy and in accord with ruling universalistic value systems' (Habermas, 1976: 58). In liberal democracies, loyalty and consent is established through the electoral and parliamentary systems that allow citizens a say in how they are governed by voting on measures (for example, referenda) and/or by electing representatives who serve on their behalf. In authoritarian systems, rule is achieved by coercive mechanisms and by the generation of fear and use of force.

It is thus not surprising that the 'crisis' of COVID-19 has been constructed and responded to differently in China than in countries with liberal democracies. China, from the outset of the pandemic, took a 'zero-tolerance' approach to COVID-19, backed by harsh lockdowns, close surveillance, and punitive measures – measures that are congruent with its authoritarian mode of governance that would be intolerable in liberal capitalist democracies, and indeed eventually proved to be so in China. The relaxation of restrictions in that country in December 2022 has been attributed in part to widespread protests which threatened to undermine the legitimacy of Communist rule (Davis, 2022).

The ideas developed by crisis theorists, I suggest, are useful in helping make sense of authorities' responses to the onset of the COVID-19 pandemic (and potentially other pandemics), which have involved efforts to establish preferred framings of the disease and its origins, risks of infection (and reinfection), and measures for its containment, among other issues – using particular imagery and language (for example, of science, of war) to engender support for risk management measures. These crisis responses have exposed inherent tensions in the simultaneous pursuit of economic and social goals in contemporary capitalist societies and the crucial role played by the state in stabilizing economies and legitimizing the social order.

To gain insight into the workings of the crisis politics of COVID-19, it is instructive to revisit the start of the pandemic, which proved crucial for subsequent responses.

Analysing a critical juncture of COVID-19

In the following paragraphs I focus on authorities' responses to the COVID-19 pandemic in the US and UK at a critical juncture in its unfolding and official definition as 'a crisis' warranting concerted action; namely, from the first reports of a viral outbreak in Wuhan, China on 31 December 2019 until late March 2020, when governments implemented containment measures of varying duration, scale, and scope. The measures included, notably, lockdowns, and restrictions on social gatherings. Critical junctures are important to analyse since they have the potential to have an enduring impact on decisions and limit alternative options; they may lead to the establishment of institutions that may be difficult to shift, due to self-reinforcing processes (Capoccia and Kelemen, 2007).

As Capoccia and Kelemen explain, during such junctures 'the range of plausible options available to powerful political actors expands substantially and the consequences of their decisions for the outcome of interest are potentially much more momentous' than at other times (2007: 343). The declaration of a state of emergency ('a crisis') presented options for action that are not available to authorities at other times or are available only to a limited extent, such as restrictions on individual freedom of movement, which is an established taken-for-granted right in Western capitalist democracies. During such junctures, authorities typically seek to control the representations of an event and its significance through their public communications such as press releases.

The period of interest includes the announcement by the WHO Director-General that COVID-19 'can be characterized as a pandemic', on 11 March 2020, and covers the start of total lockdowns in the UK (23 March) and US (19–24 March, depending on the state) and measures imposed in the two to three weeks that followed. I also analysed significant subsequent related issues (for example, lockdown resistance) reported in national news media up to the end of June 2020. While the focus is on the UK and the US, many countries around the world implemented lockdowns in March 2020 (for example, France, 17 March; Australia, 23 March; Germany, 23 March; Greece, 23 March; India, 25 March) along with other containment measures. While national lockdowns occurred around the world, their duration and severity differed markedly, reflecting different national approaches. Despite both the UK and the US being liberal democracies with technologically advanced healthcare systems that were widely expected to meet the challenges posed by a pandemic, authorities in both countries struggled to articulate consistent pandemic responses. In each case, authorities sought to strike a balance between values of collective responsibility and individual freedom of action; yet countries displayed different levels of social solidarity and confidence in authorities' ability to successfully negotiate the tensions arising from the imposition of different risk management measures.

I conducted a search and analysis of documents published online by major supranational organizations (especially the UN and the WHO), and by governments and public health departments in the UK and the US. The UN, established after the Second World War, has as its stated aim to ensure peace and security, protect human rights, deliver humanitarian aid, promote sustainable development, and uphold international law (United Nations, 2020a). As the most internationally represented and powerful supranational organization in the world, it carries great authority and has played a major role in supporting national and regional public health responses to pandemics and providing recommendations and guidance on measures. The WHO works with 194 member states spanning six regions and 150 offices, representing a diverse group of global stakeholders in public health and pandemic control in general, and COVID-19 specifically, and collaborates closely with ministries of health, government agencies, and departments at national levels (World Health Organization, 2020b, 2020c).

Documents were identified through a search of web pages and online media, including major national 'opinion-leading' news sources (for example, the *New York Times*, the *Washington Post*, and the *Guardian*) which provide a wide coverage of international issues often including links to pertinent reports and related stories. I also searched public reports of issues using Newsbank, which includes a full coverage of national news coverage spanning newspaper, newswire, video, and web-only sources. For this source, I used the search thread 'Coronavirus' OR 'COVID-19' OR 'COVID19' OR 'COVID 19'. In analysing this material, I took note of key actors and events and undertook a framing analysis, which has been used productively in studies of news and other media; for example, Anderson et al (2009). As Nisbet and Mooney explain, 'Frames organize central ideas, defining a controversy to resonate with core values and assumptions. Frames pare down complex issues by giving some aspects greater emphasis. They allow citizens to rapidly identify why an issue matters, who might be responsible, and what should be done' (Nisbet and Mooney, 2007: 56).

As I explained in Chapter 1, framing is achieved through attention to certain facts, values, imagery, and arguments, and inattention to others, which may shape how information is understood and acted on. The early framing of an emerging issue or problem in news and other media has been shown to play a crucial role in establishing the terms for debate, which may be difficult to dislodge once established; consequently, contending claims-makers often vigorously compete to control their representations (Anderson et al, 2009).

The nature of the framing is likely to be especially important in the early stages of a major disruptive event, such as a pandemic, when societies are still grappling with its impacts and policy makers have yet to form firm views on the event's significance. For this study, I focused on the overall emphasis in coverage (for example, dominant themes and use of particular sources), national and supranational authorities' rationales for stated views and actions, and the use of specific imagery and language in communications that invite or limit certain emotional and/or practical responses. I also made note of issues and facts *not* widely reported that are revealed by reading beyond the sample materials (for example, predictions of a pandemic and its disruptions), whose absence has proved critical in its overall framing.

Framing 'the crisis'

As Walby argues, crises are periods when consensus is fractured, and the range of potential meanings opens up, before narrowing again as social relations become institutionalized. They constitute a moment in time when events have the potential to have a disproportionate impact (Walby, 2015: 34). During such periods, powerful groups compete to control the interpretation of events, including their causes, magnitude, and likely impacts.

The WHO first reported 'a pneumonia of unknown cause detected in Wuhan, China, on 31 December 2019', which was then followed by the declaration of a 'public health emergency of international concern' on 30 January 2020. On 11 February, the WHO named the new coronavirus, COVID-19, and then on 11 March the Director-General of WHO announced at a media briefing that 'We have ... made the assessment that COVID-19 can be characterised as a pandemic' (World Health Organization, 2021b). A contagion that was at first framed as a localized outbreak that was 'China's problem' and contained mostly or entirely within that country's borders was reframed by the WHO within a relatively short period of time (two months) as a global 'public health crisis' demanding an internationally coordinated response. However, as I will show, events did not play out as envisaged in the official communications of major supranational bodies.

'Science and solidarity'

Recognizing that the world was confronting an emergency situation, both the WHO and the UN urged countries to make a coordinated response to the pandemic, using risk management mechanisms at their disposal to control infections and minimize the economic and social damage the expected to result from containment measures, with strong appeals made to 'science and solidarity'. In their pleas for such action, both the UN's Secretary-General, António Guterres, and the WHO's Director-General, Tedros Adhanom Ghebreyesus, called for a 'commitment to science and solidarity' (United Nations, 2020b, 2020c, 2020d; World Health Organization, 2020d). Common calls of 'we're all in this together' implied a commonality of situation and purpose which was soon belied by the responses. I discuss these early pleas for unity and coordinated action in Chapter 4, where I highlight the failures of the G20 in responding to the pandemic as demanded by the supranational authorities.

As events in March 2020 soon made clear, gaining agreement among national governments and public health authorities on the measures, and establishing citizens' compliance with them, was a far from straightforward matter. From the outset of the pandemic, different claims-makers sought to control the portrayal of the pandemic, its origins, nature, risks, magnitude, impacts, and likely outcomes, making selective reference to 'science' and/or 'evidence' to support their preferred framing of issues. In some countries, including those with technologically advanced healthcare systems, authorities' lack of commitment to or incapacity to offer a concerted, consistent response proved to be catastrophic on many levels.

Crisis messaging

As Walby also notes, 'The interpretation of a crisis may under- or over-state its magnitude and impact, as well as attribute blame as to its cause' (2015: 14). In framing a crisis, narrative is all important, and defining an event as an 'emergency' (or 'catastrophe', 'disaster' or 'crisis') allows authorities to take actions that may place aside usual democratic procedures. This situation involves the exercise of extraordinary powers, as Agamben (2005) has argued. Many authorities, and notably the UN and the WHO, used messaging that highlighted the perilous situation facing the world and the need for concerted, coordinated action. In the days leading up to national lockdowns in many countries, the language of crisis and patriotism (as in calls to national duty and good citizenship) pervaded the communications of the UN and the WHO as well as some national governments. The use of military analogies served to underline the gravity of the situation and the urgent need for a globally synchronized response.

On 23 March, the day of many national lockdowns, the UN's Secretary-General convened an 'emergency virtual summit' of the G20 'to respond to the catastrophic challenges posed by the COVID-19 pandemic' and called for a 'concerted and decisive action by world leaders' (United Nations, 2020c). In his Letter to members, he emphasized the seriousness of the situation, making references to the fact that 'Even in the wealthiest countries, we see healthcare systems buckling under pressure' and that 'A recession is in prospect'. The Letter highlighted uncertainties about the length the pandemic would last, and the damage it would produce to economies and livelihoods. Despite these uncertainties, the Letter noted that the pandemic 'will require a response like none before – *a "war-time" plan in times of human crisis*' (United Nations, 2020c, emphasis added).

The Letter called for a 'coordinated suppression of the virus', guided by the WHO, and stated that G20 countries should develop an 'articulated response mechanism' to stop transmission: 'test, trace, quarantine, treat the sick and coordinate measures to restrict movement and contact' (United Nations, 2020c). The WHO also designed a web page offering technical guidance on communication about COVID-19 which included, among other items, an 'SMS message library' which is 'intended to be locally adapted and delivered to the general public in countries around the world' (World Health Organization, 2020e). The WHO called upon 'all telecommunication companies worldwide to support the delivery of these messages and unleash the power of communication technology to save lives from COVID-19' (World Health Organization, 2020e).

The use of military language, including references to 'war' and 'fighting' and to 'saving lives' is rife in UN publications on COVID-19, as well as the communications of authorities in the UK and US, which suggests the need for countries to rally in defence against a common enemy. In the UK, on 18 March, for example, Prime Minister Boris Johnson commented at a coronavirus press conference that 'We must act like any wartime government and do whatever it takes to support our economy' (Rawlinson, 2020). In the US, New York Governor Andrew Cuomo declared at a coronavirus press conference on 30 March that 'This is a war' and that 'in this battle, the troops are our healthcare professionals' (Melendez, 2020).

The use of military metaphors can be traced back to the 17th century; however, since the beginning of the 20th century they have been widely used in medicine and public health, in a series of 'wars' against, most notably, acute infectious disease, tuberculosis, cancer, AIDS, diabetes, and obesity (Nie et al, 2016: 3). The adoption of military language – such as 'wartime plan', 'battle', and 'troops' – has served to catalyse the public to act against a common 'enemy'. However, such language can be problematic, especially in emerging health-related fields (Nie et al, 2016). The use of such language can reinforce the biomedical model by giving undue emphasis to bio-physical aspects while downplaying or ignoring the socio-political, psychological, communal dimensions of illness and healing (2016: 5–6). It can contribute to the stigmatization of certain groups or, as in the case of COVID-19, whole communities or nations which are seen as (potential) sources of disease. Indeed, this was evident very early in the pandemic when, immediately following the US President Trump's reference to 'the China virus', the number of coronavirus-related tweets with anti-Asian hashtags rose significantly (Hswen et al, 2021). The use of the hero metaphor, also widely used to described healthcare workers (often described as 'healthcare heroes') during COVID-19, notably in news media reporting, conveys expectations of endurance and 'infinite care potential', which overlooks inequities in health-work practices, the complex array of tasks workers undertake, and the 'the human toll of a system on the verge of collapse' (Navuluri et al, 2021: 211–212).

National responses

In the event, at the national level, responses to the WHO's and UN's calls were far from coordinated and military like; indeed, they were highly variable and, in the UK and the US, uncoordinated and at times chaotic. The WHO's messages were clouded by adverse media coverage of its response, and some member countries reported disregard of its guidance, with the UK and the US being explicitly mentioned (Buranyi, 2020a). The UK, widely acclaimed as the world leader in pandemic preparedness, was seemingly complacent, at least to begin with, with one commentator suggesting that the government apparently assumed that its scientists and 'evidence-based' advisors would help it avoid the kind of scenario that was unfolding in Italy at the time (van Leeuwen, 2020: 18).

On 12 March, the UK Prime Minister, Boris Johnson, was reported to have confirmed that the most stringent measures applied in other countries would *not* be applied in the UK – although within a week this advice would change, with schools closing on 18 March, and then on 23 March the government announcing the lockdown (Taylor, 2020). In 2018, the UK Government had published a biological security strategy to address the threat of pandemics; however, as a former government chief scientific advisor explained, this 'was not properly implemented' due to a 'lack of resources' and the 'distractions of Brexit and then the December general election' (Carrington, 2020).

As recently as 2017, the UK Government had produced its National Risk Register of Civil Emergencies, which noted that 'there is a high probability of a flu pandemic occurring' but added that 'the likelihood of an emerging infectious disease spreading within the UK is assessed to be lower than that of a flu pandemic' (Cabinet Office, 2017: 34). And, in 2016, the Government undertook a simulation of a flu outbreak (Exercise Cygnus) which war-gamed the UK's pandemic preparedness. The report, which the Government refused to publish, but which was leaked to the UK's *Guardian*, noted that 'The UK's preparedness and response, in terms of its plans, policies and capability, is currently not sufficient to cope with the extreme demands of a severe pandemic that will have a nationwide impact across all sectors' (Dyer, 2020).

Factors mentioned in media coverage as contributing to the UK's lack of preparedness include the passage of time since a major public health threat, combined with impacts of post-global financial crisis austerity measures and the disregard of early advice of the WHO (Wenham, 2020). One scientific assessment attributed the UK's delayed and ineffective response to an absence of appropriate expertise in the UK's scientific advisory groups required to develop and implement appropriate measures, combined with the deterioration of public health during years of austerity (Scally et al, 2020).

The US was also ill prepared for the pandemic and its responses were often delayed or disorganized during February and early March 2020, according to some assessments (for example, Zurcher, 2020). Individual states in the US eventually began moving towards lockdown, albeit at an inconsistent pace, which allowed time for the virus to spread (Zurcher, 2020). In the view of the US's top infectious diseases expert, Anthony S. Fauci, rising cases of infections were attributed to the lockdown measures being more lenient than in some European countries that had managed to stem infections (Bellware et al, 2020). Other commentators drew attention to the federal governments' failures to track outbreaks and implement testing plans and to support and fund states' and local jurisdictions' efforts to prepare for a pandemic, including stockpiling necessary equipment and supplies (Levine, 2020).

The US's federated system meant that local and state public health agencies operated in a relatively autonomous way, leading them to interpret the guidelines of the Centers for Disease Control and Prevention (CDC) differently, resulting in a 'patchwork of responses around the country' (Lewis, 2020). These delayed, uncoordinated, and ill-prepared responses provided scope for different stakeholders to exploit uncertainties by imposing their own interpretations of the pandemic and its risks, which presented an acute problem for authorities in their efforts to control the framing of the pandemic. This framing involved a definition of the pandemic that called for resolution through an 'arsenal' of technologies and techniques of risk governance.

Failures of risk governance

These uncoordinated and often delayed responses revealed the limitations and failures of 'risk governance', which dominates decision making in public health and other fields of contemporary policy. In Ortwin Renn's influential conception, which I discussed in Chapter 1, risk governance encompasses more than those elements of risk analysis of primary concern to public health and medical experts (that is, risk assessment, risk analysis, and risk communication) and requires consideration of the institutional, legal, political, social, and economic aspects of the context within which risk is evaluated and the role of various stakeholders and actors who represent them (Renn, 2008b: 8–9). Risk governance is 'a multifaceted and multi-actor risk process' and is of particular importance in, albeit not restricted to, situations where the nature of risk requires coordination between and the collaboration of different stakeholders, which are likely to include government agencies, industries, science and academia, and civil society actors (for example, non-governmental organizations) (2008b: 9). Risk governance is based on a rationality of governance that assumes the future is predictable and controllable. Being calculable, science-based risk assessment provides the basis for preventive ('risk management') actions. Yet the knowledge claims of science are rarely stable and uncontested, and the uncertainties of the coronavirus (COVID-19) disease provided considerable scope for the dissemination of non-science-based interpretations of its origins, causes, risks, and cures – as I will explain shortly.

Failures of risk governance occur for a variety of reasons, including a lack of consensus on 'risk', its source/s, character, magnitude, and approaches to its mitigation and management, making coordinated and collaborative action difficult. These failures had been evident in many countries during the COVID-19 pandemic, but perhaps most evidently in the UK and the US, which showed a lack of agreement among decision makers on the nature and magnitude of the threat and the measures. As Renn argues, in responding to risk, societies are selective as to what is worth considering and what should be ignored, and this selection is guided by cultural values (for example, the sanctity of human life), institutional and financial resources (for example, decisions by governments to invest in early-warning systems against high-consequence events), and systematic reasoning (for example, use of probability theory to distinguish the likely outcome of events or methods to estimate potential damage or distribution of hazards in place and time), among other factors (2008b: 3). Each country and risk management domain may create distinct ways of dealing with risk, or what Renn calls 'governmental styles', for example, consensual and adversarial – although these styles have become similar as globalization has proceeded (Renn, 2008b: 358–359).

These different styles are brought into sharp relief during a major risk event such as a financial crisis or pandemic when authorities are compelled to prioritize actions in order to manage the impacts. This is likely to be especially the case in situations of great uncertainty, as with the novel coronavirus (COVID-19) disease, for which there is no established precedent in how best to respond. The novel character of the disease meant that policy responses have been not only uncertain but more contested and ambiguous than with other events such as earthquakes and bushfires or other, more familiar infectious diseases, such as cholera, for which there are established operating practices (Capano et al, 2020: 288). As Capano et al explain, the high level of uncertainty surrounding COVID-19 'opened up more discretion' regarding the policy and official responses to expert advice than is the case with crises whose causes and outcomes are better known (2020: 289). Because there was no precedent for policy and action there was greater scope for national governments and public health authorities to impose their own interpretations of the pandemic and the required measures for its containment.

This high degree of discretion may be highly consequential, since, as Sylvia Walby (2015) explains, a crisis could be a 'critical turning point' or 'tipping point' in a system in which there is potential change. As she argues, a crisis is 'an event that has the potential to cause a large detrimental change to the social system and in which there is lack of proportionality between cause and consequence'; it is 'both "real", in the sense of actual changes in social processes, and socially constructed, in the sense that different interpretations of the crisis have implications for its outcome' (2015: 14). There is a short period in which interpretations and related actions may be especially important to future outcomes, although this does not mean that change will in fact occur (2015: 27, 33).

The significant discretion enabled by the disruptions and uncertainties of pandemic crisis conditions allowed much scope for different claims-makers, involving both state and non-state actors, to reframe the pandemic. As I argued in Chapter 1, public discourse on COVID-19 uncovered many different pandemic realities, as evident in communications on infection risk, vaccine effectiveness and safety, the value of different measures, among other issues. Official communications, which were often unclear and inconsistent, vied with extensive news media coverage of variable-quality and online content created by different communities, who created and shared their own pandemic narratives.

The 'infodemic' and social media

On 15 February 2020, one month before the official declaration of the COVID-19 pandemic, Tedros Adhanom Ghebreyesus, WHO's Director-General, announced at the Munich Security Conference 'We're not just fighting a pandemic; we're fighting an infodemic' (Zarocostas, 2020a). As medical and public health experts argued from early in the COVID-19 pandemic, the infodemic, comprising fake news, misinformation, and conspiracy theories, is prevalent in the age of social media and both creates confusion and undermines trust in health institutions and programmes (Zarocostas, 2020a, 2020b). Again, military metaphors were evident in these official communications in references to 'fighting an infodemic', a term which posits information itself as an 'enemy' along with the coronavirus.

A report published on 29 February 2020 in *The Lancet* noted that while it was recognized that every outbreak stretching back to the Middle Ages was accompanied by an avalanche of information, along with misinformation and rumours, social media had allowed information to be amplified and to travel faster and further, 'like the viruses that travel with people and go faster and further' (Zarocostas, 2020a). In another article, the same author explained, 'Misinformation confuses by diluting the pool of legitimate information' (Zarocostas, 2020b). Official concern about the infodemic at this time revealed concern that non-science-based information may 'go viral' and 'infect' their science-based messages of risk management and mitigation.

Consistent with Becker's view on the chaotic paradigm and challenge to the notion of predictability and linearity has been the widespread questioning of the claims of science, including about 'risk', which was evident prior to the pandemic and threatened trust in the credibility and legitimacy of established expertise. Contemporary public health is largely a discourse of risk and risk management. The cornerstone of public health is epidemiology, which focuses on identifying the 'factors of risk' for disease and targets for preventive action. Being probabilistic, risk is, by definition, calculable, predictable, and hence (in theory) governable. Yet, even during the early months of its unfolding the COVID-19 pandemic provided ample evidence of the limits and failures of risk governance – in communicating on the pandemic and control measures and, in many cases, establishing community consent for often highly restrictive measures – underlining the inescapably socio-political dimensions of risk definition and responses (Chapter 3). As became increasingly clear, appeals to abstract notions such as 'the common good' and 'solidarity' were found wanting. Some groups were using platforms such as Facebook and Instagram to claim that the coronavirus pandemic is a 'manufactured crisis designed to allow a cabal of billionaires, Big Pharma and global bureaucrats ... to use a vaccine as a ruse to implant microchips in people' (Bogle et al, 2020).

A study which explored the accuracy of YouTube videos on COVID-19, undertaken very early in the pandemic (on 21 March 2020), found that of 150 videos screened, about 27 per cent were assessed as containing 'non-factual information', totalling 62,042,609 views (Oi-Yee Lee et al, 2020). YouTube, Vimeo, and Facebook were used for 'an extremely long "trailer"' to promote a full-length film 'Plandemic', including interviews with a well-known figure in the anti-vaccination movement, which 'racked up "1.8 million views, including 17,000 comments and nearly 150,000 shares" according to a Digital Trends report, before the platforms removed the video' (Andrews, 2020). Other research identified a surge of bot-generated Twitter accounts discussing COVID-19, many created in February 2020, which had been amplifying 'misinformation', including 'false medical advice, conspiracy theories about the origin of the virus, and pushes to end stay-at-home orders and reopen America' (Hao, 2020).

Another study, of English-language communications on Twitter, Facebook, and YouTube within the UK and US contexts, undertaken in September 2020, identified 'an emerging trend ... relating to the term "climate lockdown" ... driven primarily by climate sceptics, who claimed that the COVID-19 pandemic was merely a precursor to future "green tyranny", and that both governments and global elites would curtail civil liberties under the pretext of climate change' (Maharasigam-Shah and Vaux, 2021: 4). According to the researchers, climate denialists were attempting to 'generate hype around "climate lockdown" [on social media] from as early as March 2020 ... but only gained traction following "poorly thought-out headlines and posts from mainstream institutions" which "provided grist for a reactionary media ecosystem"'. The authors concluded that the 'climate lockdown' narrative is 'not a story of outsider threats to popular discourse, but a lesson in how any message can be weaponised by those intent on harm – whether to profit from disinformation and manufactured outrage, to fuel mistrust in institutions, or to confirm existing biases about certain groups and causes' (Maharasigam-Shah and Vaux, 2021: 4).

As is common with credibility contests, authorities sought to counter these claims and bolster their epistemic authority by delineating the boundary between 'science-based' or 'factual' information and advice, on the one hand, and untrue, untrustworthy, non-science-based claims, on the other (Gieryn, 1999). In undertaking this 'boundary work', authorities claimed the moral authority to impose restrictive measures and to dismiss rival claims as 'misinformation' or 'rumours', using a variety of online resources. A notable example is the European Union's Disinfo Lab, which assembled an online repository of 'essential resources for those interested in tackling the coronavirus infodemic', with 'information on what the online platforms are doing to combat coronavirus mis- and disinformation' (European Union, 2020a, 2020b).

On 8 February 2020, the Director-General of WHO released a media briefing, emphasizing the importance of 'facts' and the 'confusion' and 'fear' generated by 'misinformation' (World Health Organization, 2020f). The media briefing then went on to outline the WHO's strategy, including working with a range of digital and social media companies and asking them to 'filter out false information and promote accurate information from credible sources like WHO, CDC and others', and 'connecting with influencers, through Instagram and YouTube, among others, to help spread factual messages to their followers, with a focus on the Asia-Pacific region'. It concluded: 'In essence, to fight the flood of misinformation, we are building a band of truth-tellers that disperse fact and debunk myths' (World Health Organization, 2020f). In May 2020, it was announced that the WHO could collaborate with the UK to run 'an awareness campaign about the risks of incorrect and false information regarding the Coronavirus pandemic', which would encourage citizens to 'double check information with trusted sources such as the WHO and national health authorities' (World Health Organization, 2020g).

In the UK it was announced in late March 2020 that a 'rapid response unit with the Cabinet Office' would be 'working with social media firms to remove fake news and harmful comment' (BBC News, 2020). The US's CDC established a 'Stop the Rumours' web page and a list of what is known about COVID-19 under a heading 'Know the facts about coronavirus disease 2019' (CDC, 2020). It also produced a web page, 'Buyer beware: how to avoid scams & misinformation during COVID-19', which included various warnings about 'scams and price gouging' during the pandemic and how to 'stay informed' and 'help put a stop to misinformation'.

In the event, as these measures were rolled out across the world, the limitations of appeals to 'facts' or 'science-based' information became evident. There were reports of people disregarding official advice on social distancing. In the UK and the US, citizens were observed to be ignoring social distancing advice (Nair, 2020; Schnell, 2020), with one survey revealing that younger people in the UK were being less compliant (Malnick, 2020). However, political leaders themselves were also reported to be flouting social distancing rules through, for example, shaking hands with strangers and meeting in large groups (for example, Fandos and Sanger-Katz, 2020). The UK's Prime Minister, Boris Johnson, even boasted that he shook hands 'with everybody' at a hospital where there were confirmed coronavirus patients – two days after being advised by the government's own scientific advisory group against greetings such as shaking hands and hugging (Mason, 2020). Later, in May, the Prime Minister's Chief Advisor, Dominic Cummings also made headlines in the UK after breaching the lockdown rules by leaving London for his second home in County Durham, which was reported to have '"clearly undermined" efforts to win the public's support to fight the pandemic' (Merrick, 2020).

Lockdowns also soon met with some resistance, often on the stated grounds of economic hardship or 'the protection of civil liberties'. Some groups, such as the UK-based 'Lockdown Sceptics', whose website included the by-line 'Stay sceptical. End the lockdown. Save lives', directly challenged the basis of authorities' 'evidence-based' arguments (Lockdown Sceptics, 2020). Through April and May 2020, acts of 'lockdown resistance' were widely reported, with thousands of people sometimes taking to the streets to protest against restrictions, often in breach of social distancing guidelines (for example, Crow and Waldmeir, 2020; Hutchison, 2020).

In short, during this early, critical juncture of the pandemic, the WHO and the UN failed to achieve a coordinated response, with national political rather than scientific considerations often foregrounded in public discourse in the UK and the US, at least. In both countries, early responses were unprepared, slow, uncoordinated, and suggested complacency, with both faring poorly in the first wave of the pandemic. At the same time, both these countries and other countries had to contend with an 'infodemic' and reports of the public's (and some politicians') flouting of public health experts' advice and acts of resistance to measures, providing a narrative that was counter to the WHO's and UN's calls for a coordinated, 'military-like' response to the pandemic.

Conclusion

It will be some time before the longer-term implications of the COVID-19 crisis are known. As Walby notes, uncertainty regarding the outcome of a crisis is integral to its definition (2015: 31). However, the history of crises suggests that there are likely to be cascading economic and social impacts. In May 2023, when the COVID-19 pandemic was officially declared over, the impacts of related control measures continued to reverberate, with supply chain and labour disruptions and recessionary effects related to economic upheavals and stimulatory measures, and surging energy prices linked to the war in Ukraine, profoundly impacting countries around the world. I explore the dimensions and impacts of these disruptions and measures in Chapters 4 and 5. The pandemic made evident the deep interconnection between national economies in terms of not only the flow of goods and energy, but also the rapid circulation of online information and of data that makes this possible. It illustrated the cascading effects of crises in the contemporary period, where an initial event predisposes to other crises, including a legitimation crisis.

Referring to the global financial crisis of 2007–09, Walby observes that this event comprised 'several interlinked phases', beginning with the banking, credit, and financial crisis between 2007 and 2009, then followed by an economic crisis and attempts by governments to reduce government deficits by cutting public expenditure, then a political crisis that carried the potential to become a democratic crisis (2015: 2–3). By 2023, it was evident that many countries, notably the UK and the US, had already been massively impacted by the cascading impacts of the COVID-19 pandemic, with incalculable social and economic costs, which then created the prospect of a legitimation crisis.

In considering sociology's contributions to the study of pandemics some years before COVID-19, Dingwall et al note that emerging diseases generate instability and uncertainty and crises that 'make visible features of the social order ordinarily opaque to investigation' (2013: 168). While those in the midst of a pandemic-induced crisis may experience the disruptions as novel and unfamiliar or, as COVID-19 has been described, 'unprecedented', viewed historically these disruptions may be viewed as inevitable aspects of the dynamics of social systems that are made visible by a certain confluence of circumstances, such as a change in the biological environment. Some researchers and public commentators suggest that human-induced climate change will mean that pandemic-related disruptions will recur in the future, creating ongoing economic, political, and social dislocation (for example, Gupta et al, 2021; Carlson et al, 2022). (I elaborate on the implications of climate change for pandemic societies in Chapter 6.)

In the view of some scholars, crisis is an ever-present feature of democratic capitalist society, given the inherent disequilibrium and instability created by the conflicting regimes of market allocation or the 'free play of market forces', on the one hand, and attention to social need or entitlement, legitimized by democratic politics, on the other (see, for example, Streeck, 2016: 74–77). Given this, pandemics may simply exacerbate the 'crisis tendencies' that beset advanced capitalism and create ongoing legitimation problems for the state that cannot be resolved through market mechanisms alone (Habermas, 1976: 58). The COVID-19 pandemic has made evident the inherent tensions and contradictions of risk governance which underpins contemporary social arrangements, and much can be learnt from constructions of and responses to this crisis for likely responses to public health risk events in the future.

In the next chapter, I examine a major manifestation of pandemic crisis politics, namely epidemiological modelling. Modelling entered widespread use during COVID-19, transforming the public representation of pandemics and their management. I will explore the critical role it played during COVID-19, and how this differed from previous epidemics and pandemics, as well as the debates about the predictive value of modelling in risk mitigation strategies and the dangers that models pose.

3

Modelling Pandemics

During the first two years of COVID-19, the language of modelling permeated public discourse on the pandemic. The declaration by the WHO that the world was confronting a public health emergency served to bring to prominence the previously arcane subjects of epidemiology, virology, and biostatistics and made familiar their vocabulary, including terms such as 'reproduction number' or R_0, 'flattening the curve', 'social distancing', 'waves of infection', 'incubation period', 'herd immunity', and 'superspreading event'. At the same time, citizens were introduced to many of their practitioners who, via their frequent media appearances or writings, became celebrities, offering seer-like predictions of what lay ahead in terms of the unfolding of the pandemic if certain measures were implemented or not.

This chapter examines the critical role played by modelling in the framing of COVID-19 and considers the implications for science and the public representation of pandemics. As portrayed by epidemiologists and infectious disease experts, models offer predictions of likely futures – the number of infections, hospitalizations, and deaths, how suppression can be achieved, and the effect of different vaccination strategies and of resuming activities following an intervention based upon certain parameters, which enable policy makers to make rapid decisions (McBryde et al, 2020). However, as I argue, models do not just predict futures; they help *shape* futures in ways that are rarely acknowledged in the media and in science debates about their utility in public health. I explore how modelling came to achieve the status that it did during the COVID-19 crisis, highlighting the consequences for how pandemics are viewed and managed, how science is practised, and how scientists view their own role.

To begin, I believe it would be helpful to first clarify how scientists conceive models and modelling – and specifically epidemiological modelling – and their applications and limitations. I then discuss the role modelling played in the control of infectious diseases before COVID-19 and some of the factors that have until recently impeded its utilization in public health. This sets the scene for the next section, where I assess the impacts of digitization on how modelling was practised and represented during the pandemic. As I conclude, modelling is based on a crude utilitarian logic that has had harmful impacts on some groups during COVID-19 and contributed to the generation of civil unrest and resistance to related mitigation measures.

What are models? What is modelling?

According to Britannica (2022), 'scientific models are at best approximations of the objects and systems that they represent – they are not exact replicas', and consequently, 'scientists constantly are working to improve and refine models'. The definition notes that models are used for different purposes, including for visualizing an object or system, and for prediction of events such as 'earthquakes, tsunamis, epidemics, and similar large-scale disasters'; however, since predictive models are unable to account for all variables, 'scientists must make assumptions that can compromise the reliability of a predictive model and lead to incorrect conclusions'. As the definition elaborates, scientific modelling is 'the generation of a physical, conceptual, or mathematical representation of a real phenomenon that is difficult to observe directly' and is 'used to explain and predict the behaviour of real objects or systems and are used in different disciplines'. Finally, it acknowledges 'the fact that models generally are not complete representations' and that 'to fully understand an object or system, multiple models, each representing a part of the object or systems, are needed' (Britannica, 2022).

Given the limitations of models, scientists have long warned of the dangers of placing too much confidence in them for decision making. In the 1970s, for example, British statistician George Box wrote an article on science and statistics where he warned that the scientist must not 'fall in love with his [sic] model' (1976: 792). In Box's view, scientific learning is progressed neither through theoretical speculation nor the 'undirected accumulation of facts' but, rather, through an 'iteration between theory and practice'. 'The good scientist', he said, should 'seek out, recognize, and exploit' errors, or the discrepancies between theory and practice (1976: 791). As Box noted:

> Since all models are wrong the scientist cannot obtain a 'correct' one by excessive elaboration. On the contrary following William of Occam he should seek an economical description of natural phenomena. Just as the ability to devise simple but evocative models is the signature of the great scientist so overelaboration and overparameterization is often the mark of mediocrity. (1976: 792)

Box recognized that when scientists applied mathematics to subjects such as physics and statistics they make 'tentative assumptions about the real world' which they know to be false but 'may be useful nonetheless' (1976: 792).

Epidemiological modelling

The discussion thus far pertains to mathematical models *in general*. But what about epidemiological models and modelling? As explained by Sofonea et al (2022), this kind of modelling 'describes a set of approaches where mathematical, statistical, and computational tools are used to study the spread of communicable pathogens in host populations'. They note, epidemic models use data and hypotheses which describe demographic processes, characteristics of environments, opportunities for transmission, and the health consequences of disease. The models can be used to verify 'what if' experiments and conjectures 'based on the construction of forecasts and scenarios' and aim to provide 'a simplified representation of a real phenomenon, focusing on the subset of properties and processes considered as essential drivers'. Consequently, their construction results from structural choices founded on previous knowledge, data, and observation and 'a set of assumptions based on our understanding of the phenomenon under study' (Sofonea et al, 2022). In developing their models, scientists tend to ignore social structures, local mores, and practices, and focus instead on the contribution of human population density, or the frequency of social contact, to contagiousness (Anderson, 2021: 174–175).

As should be clear from the foregoing, models have major shortcomings and are used when it is not possible or it is impractical to *directly* measure outcomes, as would be done in the controlled setting of the laboratory or the clinical trial. They involve selection, simplification, and abstraction with the goal of creating *a representation of reality*. As such, models operate as *technologies of framing* that direct attention on some facts and ignore others and are based on assumptions that may be incorrect and misleading. As expressed by Andreas Tolk, a computer scientist, 'Modeling is a task-driven purposeful simplification and abstraction of a perception of reality' (2015: 88). In expanding on this definition, Tolk notes that this perception 'is shaped by physical–cognitive aspects and constraints, which encompasses both observable attributes available to the observer and 'the education and knowledge of the observer, their paradigms, and even knowledge of related tools associated with the tasks' (2015: 88). Simply put, modellers cannot remove their subjectivity when producing and using models, and it is inevitable that they will need to exercise judgement drawing on their prior knowledge.

Definitions make clear that uncertainty, and hence potential error, is intrinsic to all modelling. This uncertainty may take different forms, including what modellers call aleatory uncertainty, which is 'the *inherent* variation associated with the physical system or the environment under consideration', and epistemic uncertainty, which is 'a *potential* inaccuracy in any phase or activity of the modeling process that is due to lack of knowledge' (Oberkampf et al, 2002: 336; emphases in original). This definition of epistemic uncertainty includes 'potential', since inaccuracy may or may not exist (for example, in the prediction of some event) even if there is an absence of knowledge, assuming one models the phenomenon correctly. Oberkampf et al note that epistemic uncertainty may also arise due to 'incomplete information', which they argue 'can be caused by vagueness, nonspecificity, or dissonance' (Oberkampf, 2002: 336).

In their effort to account for uncertainty, which works against calculation, explanation, and prediction – and hence the value of modelling for risk management – modellers rely heavily on mathematics, especially statistics and probability theory, which largely ignore the socio–historical context that shapes views and practices.

Epidemiological modelling in historical context

As medical historian Warwick Anderson has argued, with reference to the UK's experience of tackling COVID-19, 'Mathematical modelling has only recently come to dominate the analysis of epidemics' (2021: 171). Anderson characterizes modelling as a 'crisis technology' that has been shaped by a particular set of historical and socio–political conditions. He notes that the so-called reproduction rate or R_0, which is a key parameter in epidemiological studies (and became familiar to many people during the first year of COVID-19), began to gain traction 'during Cold War enthusiasm for global disease surveillance and epidemic intelligence'. It was first used in the mathematical theory of malaria transmission in the 1950s and then by scholars at Imperial College, London and Oxford University who applied it to other infectious diseases during the 1970s (2021: 171–172). Disease modelling subsequently gained popularity during the BSE (bovine spongiform encephalopathy) or 'mad cow' crisis of the 1990s, and then the foot-and-mouth disease outbreak. Neil Ferguson and his team (then at Oxford University) began to use modelling with viral pandemics and specifically the influenza H5N1 virus which first appeared in China in 2003 and later spread among avian populations in South East Asia (2021: 174). Ferguson and his colleagues determined that 'targeted prophylaxis' together with what they now refer to as 'social distancing' should reduce R_0 to a level that would stem inflection among humans.

While disease modelling may have gained increasing acceptance in public health in the 20th century, its origins can be traced back to the 18th century. According to Fred Brauer, the first model in mathematical epidemiology was produced by Daniel Bernoulli, a Swiss mathematician, who outlined his approach in the 1760s in his study of inoculation against smallpox (2017: 114). Further contributions to the study of infectious diseases then followed in the late 19th century: firstly, William Farr's statistical study of the laws underlining the rise and fall of epidemics, published in 1840; John Snow's analysis of the temporal and special pattern of cholera cases in the 1855 London epidemic; and William Budd's 1873 study of the spread of typhoid (Brauer, 2017: 114). Of the many statistical models used today, most are based on the so-called Susceptible-Infectious-Recovered (SIR) model, originating in seminal papers by Ross in 1916 and Ross and Hudson in 1917 and the work of Kermack and McKendrick undertaken between 1927 and 1932 (Magal and Ruan, 2014; Anirudh, 2020) (see Chapter 1).

Modelling began to find increasing application in infectious disease control during the 2009 H1N1 influenza (Swine flu) epidemic. An article published in that year which reviewed the literature of influenza modelling studies considered 'how these models can provide insights into the future of the currently circulating novel strain' of the influenza virus (Coburn et al, 2009). As the authors noted, 'The first modeling paper on influenza A (H1N1) has recently been published', namely Fraser et al (2009), showing the 'R_0 for this novel strain to be between 1.4 to 1.6' which 'is on the lower end of previous values for the 1918–1919 strain … and is comparable to R_0 values estimated for seasonal strains of influenza' (Coburn et al, 2009: 5). In their article, Coburn et al also referred to public health measures applied in Mexico which 'appear to have been successful in mitigating the outbreak of H1N1' which, they said, was corroborated by earlier modelling studies which showed that 'behavioural interventions can be very effective if R_0 is below 2'. In the authors' view, these results 'indicate that it is theoretically possible to control this pandemic' with high vaccination coverage, 'at least in resource-rich countries' (Coburn et al, 2009: 5).

However, Coburn et al added that an effective control strategy identified by modelling *might not be feasible*, even if vaccinations became available by autumn of 2009 (their paper was published in summer (late June 2009)), because 'H1N1 has now been disseminated worldwide through air travel' (2009: 5). The authors argued that, consequently, it was 'necessary for resource-rich countries to share vaccines and antivirals in order to mitigate a pandemic' (2009: 5) – a concern voiced by many people during COVID-19 with the emergence of so-called vaccine nationalism. Finally, Coburn et al (2009) drew attention to the limitations of the then current models 'in identifying effective interventions for epidemics generated by strains, such as influenza A (H1N1), that emerge due to recombination of species–species strains and subsequent cross-species transmission'. They said they had identified only two studies that had modelled interventions for influenza strains that arise from cross-species transmission. These studies, they said, 'show the potential effectiveness of quarantine as a control strategy, and also the importance of simultaneously controlling influenza in the avian population' (2009: 6).

In their conclusion Coburn et al recommended that 'more biologically complex models need to be developed' which 'could assist in identifying interventions that would be effective in reducing the probability of cross-species transmission and in mitigating pandemics driven by multi-species transmission' (2009: 7).

The Ebola epidemic in 2014 gave considerable impetus to infectious disease modelling when the US Centers for Disease Control and Prevention (CDC) established the Modeling Task Force to produce estimates of various topics related to the response in West Africa and the risk of likely importation of cases into the US. The Taskforce conducted eight Ebola response modelling projects between August 2014 and July 2015 (that is, in 'real time') to explore various questions covering the three phases of the epidemic (Meltzer et al, 2016). These were, firstly, the number of cases that might occur with and without interventions; secondly, establishing resource needs for Ebola treatment units (at the peak of the pandemic); and thirdly, generating estimates of the likely number of sexually transmitted Ebola cases (as the pandemic decelerated) (Meltzer et al, 2016).

The exercise highlighted that the modelling used to inform decisions during Ebola involved limited data, a short turnaround time, and difficulties in communicating the modelling process, which included the assumptions and the interpretation of the results (Meltzer et al, 2016). During the epidemic, the lack of access to timely data proved a major impediment to making rapid responses. Throughout the response, relevant data were not always available, or the data that were available contained inconsistencies which took time and effort to resolve. Furthermore, reporting delays made it difficult to accurately calculate the incidence of Ebola.

When adequate data did exist, the lack of data-sharing agreements meant that questions arose about who owned the data and who could use them for analysis, which delayed some modelling projects (Meltzer et al, 2016: 88). Nevertheless, modelling was seen to have proved its usefulness for future responses in terms of yielding estimates and projections that public health experts could use to make key decisions about measures and resource allocations, especially in the early stages and when data are scarce. It was argued that 'Future modeling can be enhanced by planning ahead for data needs and data sharing, and by open communication among modelers, scientists and others to ensure that modeling and its limitations are more clearly understood' (2016: 85). As the Task Force's report concluded: 'Modelling is an important but underused public health tool. Prioritising modeling in future responses will take hard work, commitment, education, and an openness in public health to new disciplines and approaches' (Meltzer et al, 2016: 89).

Modelling during COVID-19

By 2020, it could no longer be said that modelling is an 'underused' public health tool. Moreover, the argument that 'modeling can be enhanced by planning ahead for data needs and data sharing, and by open communication' seems to have struck a resonant chord among public health experts and policy makers.

As noted in the previous two chapters, the responses adopted during COVID-19 were underpinned by risk mitigation and management approaches based on *probabilistic prediction* that relies on modelling drawing on large datasets. Modelling is part of the array of risk governance approaches that authorities use to assess *probable* outcomes of events and the consequences that may follow from the exercise of certain options. In epidemiology, modelling is concerned with the identification of risk across populations as the basis for strategies of harm reduction such as immunization or drug regulation (O'Malley, 2004: 22). The modelling used in contemporary public health involves computer-generated simulations to produce probable scenarios based on existing datasets that may be incomplete or include biases based on differences of ethnicity/race, socio-economic status, gender, and age (Eubanks, 2018; Noble, 2018; Stypinska, 2023).

As Shoshana Zuboff (2019) wrote before COVID-19, we are now living in an age of surveillance capitalism, a new economic order based on 'total certainty', the decline of people's sovereignty, and an unprecedented concentration of wealth, knowledge, and power, which is enabled or facilitated by the employment of digital technologies to extract and monetize data. Zuboff's observations seem prescient, given the increases in the concentration of wealth and the acceleration of economic and social inequalities, enabled or facilitated by digitization during COVID-19 (see Chapter 5). In modern societies that are reliant on computational analyses to perform critical tasks in health and welfare and other domains, many sources may be drawn on for modelling, including the collection and analysis of data collected in 'real time'.

During the pandemic, policy makers relied on models using vast quantities of data analysed in real time to make decisions on matters such as where new cases would likely surge as stay-at-home orders expired and to justify the implementation of specific strategies such as lockdowns and social distancing to stem the spread of infections. Writing in the first year of COVID-19, sociologists Tim Rhodes and Kari Lancaster (2020) observed that modelling operated as a form of 'evidence-making' that informed rapid policy decisions. Models served to 'make the pandemic big' – of a magnitude that called for urgent or dramatic attention (Rhodes and Lancaster, 2022). Modelling undertaken in January 2020 would seem to have strongly influenced the policy on border closures following early estimates of the international spread of COVID-19 (McBryde et al, 2020).

Models were also used during the pandemic to help identify the factors that generate opposition to measures and predict 'vaccine hesitancy'. Notably, the Vaccine Confidence Project (VCP), which was founded in 2010 as part of the WHO Vaccine Safety Net developed in response to purported vaccine hesitancy and misinformation on vaccination programmes, used a diagnostic tool to detect the factors that fuelled vaccine rumours, analysed the mechanisms of their generation, and assessed the potential impact. As the VCP's website explains, the initiative is more than just about raising awareness of and knowledge about vaccines and immunization; rather, 'understanding vaccine confidence means understanding the more difficult belief-based, emotional, ideological and contextual factors whose influences often live outside an immunization or even health programme but affect both confidence in and acceptance of vaccines' (Vaccine Confidence Project, 2022a).

In 2020, scientists associated with the VCP published the results from what was claimed to be the largest known vaccine confidence modelling study based on data from 290 nationally representative surveys of 284,281 adults in 149 countries published between September 2015 and 2019. The study, which explored the importance, efficacy, and safety of vaccines, found widely varying levels of public confidence in vaccines around the world, and growing wariness in countries with political instability and religious extremism which, it was argued, 'could spell trouble as officials roll out COVID-19 vaccines in the coming months once they are approved' (Van Beusekom, 2020). In 2020, the VCP website stated that the Project was 'conducting a global study to track public sentiment and emotions around current and potential measures to contain and treat COVID-19' using a combination of population surveys and social media monitoring over time and place (Vaccine Confidence Project, 2022b).

The pandemic also provided impetus for the development of the Kirby Institute's EPIWATCH, described as a 'Global Eye' which 'monitors signals from around the world 24/7 through its AI-driven data collection including news chats, social platforms and medical reporting using over 41 global languages including the major languages in Asia' (EPIWATCH, 2022). While EPIWATCH was launched in 2016, the initiative gained momentum with infusion of funds ($800,000) from Australia's Medical Research Future Fund in 2021 to develop this 'epidemic observatory ... into a fully automated intelligent system with full language and GIS [geographic information system] capability, with a framework for seamless integration of risk analysis tools to enable rapid, early detection of serious epidemics' (EPIWATCH,

2022). According to the Kirby Institute's web page, the award 'could be a game changer in health security'.

> Researchers say if the COVID-19 pandemic had been detected early in its genesis before it spread beyond Wuhan, it could have been stamped out entirely and the pandemic prevented. Currently, the public health system relies on doctors or laboratories to report epidemics, which is a passive and untimely system. But there is a vast array of un-curated, open-source data, such as social media and news reports, which capture the concerns and discussions of the community. Algorithms and AI technology can be used to make sense of, or 'mine' this data, to reveal potential early signals of epidemics, prior to official detection by health authorities. To date however, use of this technology for early detection has not been a focus of pandemic planning. (UNSW, 2021)

The evidence-making of modelling has been based on classifications produced by individual researchers and research teams, governments, organizations, and businesses, purportedly to help understand the nature and implications of COVID-19, to track its progress, and to help manage and mitigate the risks (including potential 'hesitancy' or resistance to measures) and thus help stem infections and save lives. Crisis conditions served to focus public attention on the host of potential factors deemed liable to produce harm – the so-called 'factors of risk' (Castel, 1991).

Modelling the factors of risk was not just of concern to health authorities and scientists. Soon after COVID-19 was declared to be a public health emergency, both large and small businesses demanded 'a road map' or 'path out' of lockdowns and other restrictive measures which, they argued, were needed to make investment and staffing decisions. They looked to policy makers who rely on models so that they could understand what was at stake in the adoption of the different proposed options. The mathematized form of modelling promised 'security in numbers' even if, as has so often happened, credentialed experts and decision makers disagreed on their interpretation.

Bloomberg, a privately held financial, software, data, and media company owned in part by Merrill (previously Merrill Lynch), an investment and wealth management division of Bank of America, developed a 'Covid resilience ranking' that identified 'The best and worst places to be as variants outrace vaccinations' (Chang et al, 2021). According to a separate website which explains the methodology, 'the Covid resilience ranking scores the largest 53 economies on their success at containing the virus with the least amount of social and economic disruption', taking into account 'the progress of normalization in global travel routes and flight capacity' (Chang et al, 2021). The economies chosen, it is explained, were those valued at more than $200 billion before the pandemic. These rankings were constantly updated according to how countries were performing as rates of testing, vaccination, and other pandemic measures – including restrictions on travellers from COVID 'hot spots' – changed. They provided a kind of score card of how countries were performing in the 'fight' against the virus. It included information on the percentage of people vaccinated in the countries surveyed, their 'lockdown severity', flight capacity, and the number of 'vaccination travel routes' (Bloomberg, 2021).

Countries were given a 'resilience score' and ranked, with details of previous scores, to indicate whether they are getting 'worse' or 'better'. In September 2021, when I undertook a review of these scores, some countries, namely the US, UK and Israel, were identified as travelling in the 'wrong direction' in that 'Covid cases have surged anew in vaccine front-runners'. It also identified 'Notable movers', countries that were either moving up the 'rungs', as they expanded their travel access to vaccinated foreigners and/or eased restrictions, or 'down' as new infections rebounded, and restrictions were reinstated. The language used in this ranking system revealed the biases of this classificatory system, with references made to countries performing 'worse' or 'better' and moving 'up' and 'down' the resilience rank, implicitly conveying the desirability of certain countries and the dangers for countries and regions being ranked poorly, which would likely be significant for investors who would have been those most likely to be interested in this kind of information.

The authors of the blog acknowledged that 'Bloomberg's Covid Resilience Ranking provides a snapshot of how the pandemic was playing out in 53 major economies right now' and that the shifts in economies' performance towards 'opening up and the revival of global travel ... provide a window into how these economies' fortunes may shift in the future as places exit the pandemic at different speeds' (Chang et al, 2021). However, the creation of such a ranking, which implies that countries were in a 'race to open up', denies the inequalities between nations in access to the vaccines and resources that worked against them competing on equal terms.

UK debates on modelling during COVID-19

While modelling achieved a pre-eminent status during COVID-19, it did not always inform the policy decisions of national governments. In the UK, for example, modelling had limited impact on decision making to begin with and only later became an essential guide to policy when the pandemic was reframed in the media (Englemann et al, 2023). UK modellers (including Neil Ferguson, based at Imperial College, London) at first promoted the strategy of 'mitigation' rather than 'suppression' – to lessen R_0 and 'flatten the curve' – with the goal of building herd immunity and thus limiting or stopping transmission (Anderson, 2021: 175). But he later changed his view to support suppression, urging a lockdown until a vaccine could be distributed, after further modelling from Imperial College and observations of the high mortality rates and impositions on intensive care facilities in northern Italy showed that 'hundreds of thousands of deaths would still occur and the National Health Service would collapse – long before any herd immunity was attained' (2021: 175–176).

According to Englemann et al (2023), this shift in stance in the UK evidently owes much to changes in both policy and media framing of the pandemic in March 2020 (Englemann et al, 2023). That is, whereas early media coverage of and policy debate about the pandemic was surrounded with scepticism and then speculation about the pandemic (*whether* Britain would be affected), accompanied by policy caution, this shifted to talking about *when* it would occur along with more concerted policy action, including deferral to modellers like Ferguson in the first two weeks of March 2020 (Englemann et al, 2023: 128–129). As Englemann et al note: 'The shift was prompted not just by reports of the rapid progress of the disease in Italy and other countries, but also by a growing conviction that it was only a matter of time before it took root in the UK' (2023: 129).

The pandemic soon revealed that scientists disagreed among themselves about the 'best' models to use and whether reliance on modelling might have unintended and even harmful consequences. Anderson (2021) observes various experts' uneasiness with the use of limited parameters and the simplifications used in modelling developed by Imperial College, London (discussed in the following section), which was seen as outdated and sociologically naïve in failing to account for social complexity and heterogeneity. As he notes, experts at the time identified various flaws in the models and the limited perspective offered by mathematical modelling (2021: 176–177). Some questioned the way that data were used in models to support lockdowns and other measures while ignoring uncertainty and the margin of error.

For example, the UK Statistics Authority questioned the way the government used statistics in models to justify the national lockdown that projected 'there would be 1,000 deaths a day by the end of October [2021] when the average was actually four times less than that – a fact that was known at the time of Saturday's TV briefing' (BBC News, 2021b). In this case, it was reported that 'the model had already been updated to predict a lower estimate, but this was not used in the briefing fronted by chief scientific advisor Sir Patrick Vallance and chief medical officer Prof Chris Whitty, alongside Prime Minister Boris Johnson'. The report added: 'It is understood the graph was used by the two senior advisors in meetings last week where the decision to impose a national lockdown in England was made' (BBC, 2021b).

More generally, some scientists called for the development of more *country-specific* models that would pay cognisance to issues such as the economy, demographic mix, household composition, social values and norms, and so on (Squazzoni et al, 2020). There were also demands for future models to include the social and economic effects of lockdown and other measures such as mental health and interpersonal violence, the impacts of closures of mass transport and of leisure activities, and the effects of closing different kinds of institutions for various lengths of time (Colbourn, 2020). Researchers raised concerns about the pressures they were under during the 'crisis' of COVID-19 to respond to policy makers' demands for rapid advice in the absence of adequate reliable data and public expectations about the role of science, and the tensions this posed for their pursuit of the 'normal' path of scientific research and publication (Squazzoni et al, 2020). They also voiced concerns about the influence of the Imperial College COVID-19 model, 'which has contributed to reshaping the political agenda in many countries' despite having many shortcomings and lack of transparency regarding its 'model code' (Squazzoni et al, 2020).

Modelling and fear

While, as stated, modellers acknowledge that all models have shortcomings, these may be overlooked by authorities in a context of crisis-induced fear that demands rapid policy decisions. In the case of COVID-19, Carlo Caduff argues that this fear was fuelled by mathematical disease modelling, along with neoliberal policies that have created fragile systems unable to cope with crisis, media reporting which has provided a constant stream of information, and 'authoritarian longings' in democratic societies, which 'allowed governments to create states of exceptions and push political agendas' (Caduff, 2020: 480).

One of the major concerns expressed by authorities during the early phases of COVID-19 was the ability of healthcare systems to cope with a substantial spike of severe illnesses. Media coverage of overwhelmed healthcare systems in some countries, notably the US, Italy, Spain, and India in the early months of the pandemic, heightened concerns of similar scenarios in other countries. As noted, in the UK, a major public health goal during the early phases of the pandemic was to 'flatten the curve', the aim being to reduce the number of people simultaneously requiring treatment as well as the time at which this would occur – hence giving healthcare services longer to increase their capacity to treat patients by obtaining increased ICU (intensive care unit) bed availability and obtain equipment such as ventilators (Thompson, 2020). It was argued that 'the SIR model can be used to demonstrate the principle that a reduction in transmission can delay, and reduce the height of, the epidemic peak' (Thompson, 2020). The model is claimed to be reasonably predictive, but its limitations have long been acknowledged and soon became evident with the onset of COVID-19.

As I mentioned, experts have also acknowledged the dangers of relying on models alone for epidemic or pandemic measures. In early 2020, for example, in the context of COVID-19 lockdowns, Thompson (2020) wrote: 'Mathematical models are a key tool for guiding public health measures, and outputs from epidemiological modelling analyses should be considered alongside numerous factors (such as potential economic and mental health effects of interventions) when deciding how to intervene.' As Thompson commented, 'Perfect data are not available, so modelling requires assumptions ... [however,] despite unavoidable uncertainties, models can demonstrate important principles about outbreaks and determine which interventions are most likely to reduce case numbers effectively.' He concluded: 'Models demonstrated the need for the current lockdown [in the UK in 2020], and modelling must remain a key tool for informing policy as the lockdown in the UK is relaxed' (Thompson, 2020).

Influential modellers and models

During COVID-19, some modellers and/or their research groups achieved a high media profile and had great influence on public debates and policy, including regarding the best models to use. Some achieved a celebrity-like status, underlining the profile and influence that some experts may achieve during a crisis. Modellers with established reputations or who worked in research institutions headed by individuals with a high profile appeared regularly on television and other news media and were often called upon by governments for policy advice. In Australia, for example, the Doherty Institute, headed by Peter Doherty, who won the 1996 Nobel Prize in Physiology or Medicine for his research on how the immune system recognizes virus-infected cells, was commissioned by the Australian Government 'to advise on the Plan to transition Australia's National COVID Response' (Doherty Institute, 2022). Doherty's modelling played a significant role in decisions about strict lockdowns in Australia which were implemented to control infections in the country until widespread vaccination coverage in the population had been achieved (Doherty Institute, 2021). In the UK, a model developed by Neil Ferguson's group at Imperial College, London (Ferguson et al, 2020) is claimed to have informed social distancing measures in the UK and other countries to stem the COVID-19 spread (Enserink and Kupferschmidt, 2020; Panovska-Griffiths, 2020).

However, while certain individuals or research groups may have possessed this public profile and influence, this does not mean that their models were necessarily accepted by other experts. Some scientists raised concerns about the robustness of the model used by Ferguson's group, given that the dataset used was 'only days, possibly a couple of months, long' and that the model had to be 'drastically revised' from initial predictions of 500,000 deaths from severe infection to 20,000 deaths, using different reproduction rates (Panovska-Griffiths, 2020). Moreover, as Panovska-Griffiths (2020) argued in 2020, a different mathematical model used by a group at Oxford University which suggested that 'ongoing epidemics in the UK ... started at least a month before the first reported death' led others to wonder 'whether the seemingly different conclusions drawn exposed problems with using models for infectious diseases transmission as key drivers of policy decision making'.

Panovska-Griffiths notes that these models used different data and neither of them could answer all the questions. Whereas Ferguson's group used a model (a stochastic individual based model) that 'considers the infectiousness of each individual within the population as a function of the number of contacts within the household, work/study place and random contacts', the Oxford Model (developed by Sunetra Gupta's group) used the classic SIR model that averages the infectiousness across the population (Panovska-Griffiths, 2020). Each model suggested different strategies: either the use of suppressing interventions as key to 'flattening the epidemic curve' or undertaking large-scale antibody testing (on the assumptions that a large proportion of the population may have already been infected), respectively. However, as Panovska-Griffiths (2020) argues, used together the two models contributed to creating 'the full picture of how best to tackle COVID-19 spread'. Different models are suited to different contexts. As Panovska-Griffiths explains:

> Both types of models have been used historically across different infectious diseases and both have advantages and disadvantages, with the modelling approach chosen often based on the preference of the modeller. Under exactly the same conditions, i.e. same datasets, same parameters, using same numerical software for simulations, they ought to converge to one another. They may not, as is the case for the Imperial and Oxford models, when they use different data.

Some scientists argued that the make-up of the population affected the utility of modelling. In a small country like the Netherlands, compartment models, which assume that the population is homogeneously mixed, may be a reasonable assumption, but in larger, heterogeneous countries, such as the US or all of Europe, modelling that simulates the day-to-day interactions of millions of individuals is deemed more useful (Enserink and Kupferschmidt, 2020). As Enserink and Kupferschmidt (2020) observed, the WHO organized regular meetings of COVID-19 modellers to compare different strategies and outcomes, with the aim of reducing discrepancies between models and thus assisting policy making.

In March 2020, at the onset of COVID-19, Enserink and Kupferschmidt (2020) wrote:

> There's … at lot that models don't capture. They cannot anticipate, say, the development of a faster, easier test to identify and isolate infected people or an effective antiviral that reduces the need for hospital beds. … Nor do most models factor in the anguish of social distancing, or whether the public obeys orders to stay home. … Long lockdowns to slow a disease can also have catastrophic economic impacts that may themselves affect public health.

During COVID-19, there were many debates among scientists about the utility of different models and the impacts of modelling in general. Some scientists expressed concerns about over-reliance on modelling for decisions such as lockdowns – notably the signatories of the Great Barrington Declaration.

The Great Barrington Declaration

In October 2020, three scientists – Dr Jay Bhattacharya, a public health policy scholar from Stanford University, Dr Sunetra Gupta, an epidemiologist at Oxford University, and Dr Martin Kulldorff, a biostatistician and epidemiologist based at Harvard Medical School – wrote the Great Barrington Declaration, which raised 'special concerns about how the current COVID-19 strategies are forcing our children, the working class and the poor to carry the heaviest burden' (Great Barrington Declaration, 2022a). Purportedly 'written for the public, fellow scientists, and government officials', the Declaration was sent to scientific colleagues and released to the public on 5 October 2020.

The website established to promote the Declaration's aims included an online form which supporters could sign – which, as explained, had attracted some 'fake signatures', but 'less than 1% of the total', most of which were reportedly removed. The Declaration itself describes the 'devastating effects on short and long-term public health' of the lockdowns then in place, which were said to include 'lower childhood vaccination rates, worsening disease outcomes, fewer cancer screenings and deteriorating mental health – leading to greater excess mortality in years to come, with the working class and younger members of society carrying the heaviest burden'. It adds, 'Keeping students out of school is a grave injustice', noting that the old and infirm are much more vulnerable to death from COVID-19 ('more than a thousand-fold higher') than the young and that for children 'COVID-19 is less dangerous than many other harms, including influenza' (Great Barrington Declaration, 2022b).

Consequently, the authors argue for 'the most compassionate approach', which is a 'Focused Protection' of the vulnerable until herd immunity is reached, allowing those 'who are at minimal risk of death to live their lives normally to build up immunity to the virus through natural infection, while better protecting those who are at a higher risk' (Great Barrington Declaration, 2022b). By late September 2022, the Declaration had attracted nearly 933,000 signatures from countries around the world, with the 'signature map' showing that majority were from North America, the UK, and Europe (Great Barrington Declaration, 2022c).

Many scientists also raised concerns about modelling in articles written at various stages during the pandemic. In 2020, Anirudh, who examined the challenges of epidemiological modelling given the high level of uncertainty in each of the models used to predict the spread, peak, and reduction of COVID-19 cases, warned of the harms caused by 'over-reacting to a particular model' (2020: 372). One of the 'challenges', the author noted, was that 'The simplistic models like SIR … are struggling due to not accounting the impact of super spreaders, public places such as schools, public transport, crowded workplaces apart from delay in providing medical attention etc.' (2020: 372) Crucially, the SIR model assumes that the population is 'fully mixed'; that is, that all individuals have the same probability of contact with others, which is rarely if ever the case (Magal and Ruan, 2014: 27).

Regardless of scientists' stances on the aforementioned issues, there is no doubt that COVID-19 changed public representations of modelling and pandemics.

COVID-19's impact on public representations of modelling

Writing in 2020, Rhodes and his colleagues observed:

> Modelling nowcasts, forecasts and 'coronacasts' are featuring heavily in media portrayals and public engagements in relation to COVID-19. Modellers are communicating their maths and projections directly to publics online, and are publishing quickly, unencumbered by peer review. There is not only a race to model to produce evidence for policy, there is a thirst for *knowing maths and models* as a means to *knowing COVID-19*. (Rhodes et al, 2020: 254; emphases in original)

Some modellers are acutely aware of how their practices have shaped public representations of epidemics and pandemics. For example, Adam Kucharski, a mathematician whose research focuses on the dynamics of infectious disease, says that COVID-19 has offered a 'tragic "natural experiment" that altered people's perceptions of epidemics'. Writing an article in the UK's *Guardian* newspaper nearly a year after the onset of COVID-19 (in January 2021), Kucharski commented,

> Think back to some of the things you learned from Covid-19 in 2020: information such as 'fatality risk' and 'incubation period'; the potential for 'super-spreading events', and the fact that transmission can happen before symptoms appear. There were the suggestions in mid-January that the Covid-19 outbreak in Wuhan was much larger than initial reports suggested, and we learned how Wuhan's subsequent lockdown led to a reduction in transmission. What links these early insights? All of them involved epidemic modelling, which would become a prominent part of the Covid-19 response. (Kucharski, 2021)

As Kucharski points out, during earlier disease epidemics, such as Swine flu in 2009 and Ebola in 2014–2016, publics only learnt from the insights of modelling *after* the findings had been published in scientific papers. Soon after the onset of the COVID-19 pandemic, researchers began to build online dashboards so that citizens could track transmission levels and compare potential scenarios themselves and made preprints rapidly available. Additionally, modellers shared findings of 'real-time' modelling analysis of coronavirus variants and genetic data and case trends, providing a global pattern of how variants were spreading globally (Kucharski, 2021) – a development that I discuss later in the chapter.

As Kucharski comments, outbreak research should be 'fast, reliable and publicly available'. However, 'the pressures of real-time COVID-19 analysis' meant that many academics undertook their work in their spare time without dedicated funding, forcing them to make choices on how they should prioritize their efforts: assisting governments and health agencies, writing up articles to describe their methods, or helping others to adapt their models to different questions. Kucharski says that while these are not novel problems, the pandemic gave them new urgency. In the US, he notes, the most comprehensive COVID-19 database, the Covid Tracking Project, was run by volunteers. As the Project's website notes, 'hundreds of volunteers' have 'collected and published the most complete data about COVID-19 in the US' (COVID Tracking Project, 2022a). A link on the project's home page takes one to the list of volunteers and their expertise, including 'data tracker', 'racial data tracker', 'data quality', 'long-term care data', 'data infrastructure', 'city data', 'editorial and website', and 'reporting and outreach' (COVID Tracking Project, 2022b). In Kucharski's (2021) view, the pandemic laid bare 'inefficient and unstainable features of modelling and outbreak analysis and illustrated that there is a clear need for change'.

In their effort to quickly share their data and translate findings into innovations, many scientists published their work via preprints. This form of publication has been in existence for more than 30 years, originally to post physics papers. As Fraser (2021) notes, preprints were used to some extent to share information on the Zika and Ebola virus outbreaks; however, the COVID-19 pandemic was the first time that preprints were widely used outside specific communities to communicate during such an event. The potential impacts of this shift in scientific publishing are far-reaching and go well beyond science.

While rapid publication has enabled scientists to share findings that have assisted pandemic control efforts – in showing that pre-symptomatic transmission was occurring, and in quickly developing vaccines and life-saving therapies (dexamethasone) – they have also enabled the dissemination of fraudulent research and dubious findings (for example, on treatments) that had the potential to endanger lives and undermine confidence in public health (Watson, 2022). An example cited by Watson (2022) is a preprint published in January 2020 which suggested that the new coronavirus showed similarities to human immunodeficiency virus. While the paper was criticized and withdrawn from *bioRxiv* within 48 hours, this did little to stem the conspiracy theories that it generated. It was not long before this study was 'usurped by another questionable study of antibody seroprevalence that insinuated that SARS-CoV-2 infections were more common and less serious than feared' (Watson, 2022).

Watson highlights the dangers posed by the speed and intensity of research publication during COVID-19 – including shortcomings of the peer-review process – which may have enduring impacts. Of particular note in this regard is the publication of a French study on hydroxychloroquine (a treatment promoted by the then US President Donald Trump) with 'gross methodological shortcomings', which led to skyrocketing medical prescriptions of the anti-malarial drug, which continued to be written nine months after the publication of 'convincing evidence that it was useless for treating COVID-19', with the paper never being retracted (Watson, 2022).

Genomic sequencing

Rapid data sharing and the development of more complex models during COVID-19 was greatly assisted by rapid advances in genomic sequencing. A notable example of how genomics has been integrated into real-time public health surveillance is Nextstrain, a database of viral genomes established in 2015. On its website, Nextstrain is described as 'an open-source project to harness the scientific and public health potential of pathogen genome data'. Further, 'We provide a continually-updated view of publicly available data alongside powerful analytic and visualization tools for use by the community', the goal being 'to aid epidemiological understanding and improve outbreak response' (Nextstrain, 2022). The coordinators of Nextstrain have developed up-to-date genomic analyses of various pathogens, including West African Ebola, Avian influenza, Tuberculosis, Measles, Mumps, and SARS-CoV-2.

The analyses of Nextstrain and other genomic databases such as the COVID-19 Genomics UK Consortium (COG-UK) (launched in March 2020, soon after the UK went into its first lockdowns) and COVID-19 Host Genetics Initiative, profoundly shaped thinking about and responses to the pandemic. For example, the decision to lock down one third of the UK's population in December 2020 is claimed to have been underpinned by the findings of genomic analysis. According to John Drake, who specializes in the ecology of infectious diseases at University of Georgia, in the US, the British lockdown was evidently driven by 'the rapid increase in the number of cases represented by a specific evolutionary branch of SARS-CoV-2, the virus causing COVID-19' (Drake, 2020).

As Drake explains, the lineage was named B.1.1.7 by the COG-UK Consortium, and the findings were published in an online report in the blog virological.org, co-authored by ten experts in viral evolution (Drake, 2020). The Consortium includes 500 members comprising 'a range of scientific and business expertise in genomics, bioinformatics, operations clinical science and public health', as its 2022 website (now archived) explains (COVID-19 Genomics UK Consortium, 2022). The final report of the Consortium was published in 2022 (Marjanovic et al, 2022). Formed in March 2020, the Consortium has been ascribed a crucial role in the pandemic response by sequencing the genome of the SARS-CoV-2 virus, and by February 2021 had sequenced about half of all its genomes in the Gisaid database (Peacock, 2021) – a platform originally established in 2006 to improve the sharing of data among scientists on Avian influenza (Gisaid, 2022).

This genomic-based approach to pandemic control involved a vast network of actors and actor-networks, including national biobanks, private companies, and university researchers – providing a new collaborative model of pandemic surveillance. These initiatives have been founded on a vision of assumed common interest and altruism, driven by a sense of urgency and commitment to sharing and open communication. As explained on the home page of one of the leading collaborative ventures, the COVID-19 Host Genetics Initiative, given the urgency to obtain insights into understanding and treating the disease, it was 'critical for the scientific community to come together around this shared purpose' (COVID-19 hg, 2022).

According to its stated aim, the initiative 'brings together the human genetics community to generate, share, and analyse data to learn the genetic determinants of COVID-19 susceptibility, severity, and outcomes'. It is noted that 'Such discoveries could help to generate hypotheses for drug repurposing, identify individuals at unusually high or low risk, and contribute to global knowledge of the biology of SARS-CoV-2 infection and disease' (COVID-19 hg, 2022).

The assumed altruism that underpins the aforementioned genomic collaborations is conveyed by the statement: 'Nothing is written in stone other than we must all act together and with no personal gain or ownership of results – just rapid and immediate dissemination of the maximum possible data and information that can be responsibly released' (COVID-19 hg, 2022). An article describing the initiative and its goals describes its approach as 'inclusive, decentralized, and transparent' (The COVID-19 Host Genetics Initiative, 2020: 717). The seven 'Principles of collaboration' convey the implied trust relationship involved and underlying concerns about less collaborative practices, with five principles focusing on issues of honesty, trust, and respect for other groups' data: 'Collaborative in an environment of honesty, fairness, and trust'; 'Respect other groups' data'; 'Operate transparently with a goal of no surprises'; 'Seek permission from each group to use results prior to public release'; and 'Do not share another group's results with other parties without permission' (COVID-19 hg, 2022).

Such expressions of openness, mutual respect, and collegiality may have shaped conduct among some groups of researchers. But the pandemic also showed that professional rivalry, the imperative to be the first in announcing 'breakthroughs', and the pressure to quickly publish results, tended to work in the opposite direction. Conflicting views on certain models or modelling per se showed that scientists often disagreed on fundamental issues of critical importance to public health decision making. In the event, hopes for modelling and the risk mitigation strategy began to fade with the arrival of the Omicron variant, which cast doubt on the policies that had held sway in many countries since the beginning of the pandemic.

Fading hopes for modelling after Omicron

The arrival of the Omicron variant in the final months of 2021, which had a much higher reproduction number than the Delta variant – making it one of the most infectious diseases known – and could evade the immune response, made previous elimination and suppression strategies redundant (Vally, 2022). It was at this point that the limitations of modelling as a tool of risk mitigation became palpable. While it could be argued that strategies based on modelling scenarios saved many lives during the pandemic – a proposition difficult to prove in retrospect – by the latter months of 2021 and early 2022, it was evident that many governments had shifted their approach from suppression to 'living with the virus' – in effect acknowledging that COVID-19 was endemic in populations and could not be eliminated. Curiously, this new approach was implemented in a year when COVID-19-related deaths were at an all-time high, and higher than they were in 2020, when many countries were in lockdown. In Australia, for example, there were 14,780 COVID-19 fatalities in 2022, compared with 909 fatalities in 2020 (Grenfell, 2023).

An issue that received little attention in debates on epidemiological modelling during COVID-19 is that it is underpinned by a utilitarian approach that presumes that the benefits deriving from related measures will outweigh the costs and harms of inaction or of pursuing alternative options. This utilitarianism has informed experts' and authorities' judgements about the implementation of various pandemic measures, including lockdowns, mandated facemasks, vaccines, and tests. The principle of utilitarianism, which can be traced back to the late 18th/early 19th century English philosopher, Jeremy Bentham, takes many varieties but is characterized by the evaluation of an action in terms of its *consequences*, especially in terms of advancing the overall happiness or wellbeing of those affected by the action (Internet Encyclopedia of Philosophy, 2022).

Utilitarianism has been subject to much critique, including on the grounds of the inability to quantify, compare, or measure wellbeing or happiness and that the consequences of different adopted measures, such as those adopted during the pandemic, are unknowable (Savulescu et al, 2020). While certain measures such as lockdowns appeared to be readily accepted by many people who believed that the measures would help 'keep them safe', they also marginalized and, in some cases, stigmatized some groups who, as mentioned (Chapter 1), for various reasons did not or could not conform with the measures (Kartono et al, 2022). This also likely contributed to civil unrest and acts of 'lockdown resistance' which occurred in many countries during the pandemic (Wood et al, 2022) (see Chapter 4).

Conclusion

The COVID-19 pandemic would appear to represent a critical juncture in the history of epidemiological modelling. During this event, models operated as 'crisis technologies' that have profoundly shaped how pandemics are understood and responded to. Their use reflects the belief that, given the 'right' parameters, pandemic risks can be predicted and controlled. Yet, a pandemic is an inherently uncertain event, especially when it is caused by a novel virus calling for measures that massively disrupt societies. As I explain in the next chapter, the pandemic has shown that these measures may have unintended consequences, including the reinforcement of inequalities of many kinds.

4

Pandemic Crisis and Inequalities

In future years, the COVID-19 pandemic will likely be remembered not only for the many illnesses and deaths resulting from the disease itself but for the far-reaching social and technological changes precipitated or accelerated by the crisis. The WHO's declaration of a public health emergency in 2020 led to measures that exposed previously hidden or taken-for-granted features of the social order and produced massive disruptions to economies and societies. In particular, the crisis laid bare and reinforced inequalities of various kinds – socio-economic-, age-, and gender-based, among others – which contributed to a sharp rise in political instability and acts of civil unrest (for example, protests and riots related to restrictions) in some countries in the first one to two years of the pandemic (Institute for Economics and Peace, 2021). Moreover, the crisis revealed the failures of organizations that were established to coordinate responses to and manage economic and public health crises; notably, the G20 and the WHO.

In this chapter I examine the inequalities made evident or reinforced by the crisis, drawing on research evidence produced by academic researchers, non-governmental organizations, independent panels, the UN, the WHO, and the World Bank during or soon after the pandemic. As I argue, the crisis made clear the failure of governments to work in a coordinated way to manage the pandemic as implored by the UN and highlighted the global inequalities that worked against such a response. The events of early 2020 soon exposed fractures in the global political order, with the G20 group prioritizing their countries' own interests, which was reflected in debates about 'vaccine nationalism', which I discuss.

To begin, I revisit the UN's call to action, before turning to consider the evidence on the social and economic impacts of the crisis.

The UN's call to action

As discussed in Chapter 2, on 23 March 2020, around the time many countries implemented national lockdowns, the UN Secretary-General,

António Guterres, declared that not only were healthcare systems of the wealthiest countries under stress but 'A recession is in prospect' (United Nations, 2020c). The Secretary-General added, 'The question is: how long it will last and how much damage it will do to the productive capacities of our economies and the livelihoods of our citizens' (United Nations, 2020c). He went on to note that 'The G20 leadership has an extraordinary opportunity to step forward with a strong response package to address the various threats of COVID-19', which 'would demonstrate solidarity with the world's people, especially the most vulnerable' (United Nations, 2020c).

In Chapter 2, I referred to the role played by such crisis messaging, which serves to evoke fear and a sense of common purpose against a defined enemy – in this case what was seen as a potentially deadly disease. As noted there, fear is a feature of crises that can mobilize populations – at least where there is a widely perceived threat. In the case of COVID-19, the threat was not just the disease itself, whose character was and in 2024 remains uncertain in many respects (for example, symptom expression, mode of transmission, long-term health impacts), but the economic turmoil arising from pandemic control measures including mass unemployment and related loss of incomes and failed businesses (particularly in 2020), increased levels of poverty, and, although not explicitly articulated in mainstream media or other official communications, related social instability.

However, while in the early stages of the pandemic, many governments made statements that suggested they shared the concerns of the UN and the WHO, this was not always reflected in the responses at the time. The UN Secretary-General called for G20 leaders to 'be decisive and commensurate' and to 'inject massive resources into economies, reaching double-digit percentage points in the world's gross domestic product'. He referred to the need to 'minimize the social and economic impact of COVID-19 for everyone and stimulate a faster recovery everywhere' and warned that 'By the end of this year, the cost of this pandemic is likely to be measured in the trillions of dollars'. This called for the injection of 'massive resources into economies, reaching double-digit percentage points in the world's gross domestic product'. Moreover, it was argued, the ' "business-as-usual" economic rules and policy tools no longer apply' in these 'unprecedented times' (United Nations, 2020c).

As the UN Secretary-General noted, unlike 2008, 'this is not a banking crisis', and 'While liquidity of the financial system must be guaranteed, we need to focus on people – families, low-wage workers, small and medium enterprises and the informal sector'. The response, it was argued, called for the G20 leaders to launch a 'large-scale coordinated stimulus package in the trillions of dollars to target the direct provision of resources to businesses, workers and households in those countries unable to do so alone' (United

Nations, 2020c). As the Secretary-General noted, G20 countries account for 85 per cent of the world's gross domestic product (GDP), and therefore have 'a direct interest and critical role to play in helping developing countries cope with the crisis'. The communication went on to outline the kinds of measures required, including the use of trade credits, liquidity relief to the private and financial sectors in the developing world, along with reaffirmed commitment to 'more inclusive and sustainable models of development' (United Nations, 2020c). The UN Secretary-General's call at this time implied an altruistic intent, a concern both for citizens in poorer parts of the world and for the health of the planet, as well as a sense of common purpose (as in the common entreaty 'we're all in this together').

How then, did events unfold following the UN's call to action? Did the rich G20 countries fulfil their ascribed responsibilities?

The social and economic impacts of the crisis

Firstly, the social and economic impacts of the pandemic crisis indeed proved to be severe, especially in the first year following the declaration of the emergency, which led to border closures and lockdowns in many countries. They included disruptions to supply chains, widespread unemployment (albeit highly uneven across sectors), the closure of many small businesses, civil unrest (most evident in 'lockdown resistance' in some countries), and detrimental impacts on government budgets from both pandemic support measures and the loss of revenue from reduced income tax and other taxes. During 2020, the world's collective GDP was reported to have fallen by 3.4 per cent (about US$2 trillion), although it recovered in 2021 (Dyvik, 2024). Many economies effectively came to a standstill as companies shut their doors, staff were laid off or furloughed, and workers and students began to work at home, often for lengthy periods.

In June 2020, the World Bank issued a press release announcing that 'The swift and massive shock of the coronavirus pandemic and shutdown measures to contain it have plunged the global economy into a severe contraction.' The Bank offered the prediction that the global economy would shrink by 5.2 per cent that year, and 'That would represent the deepest recession since the Second World War, with the largest fraction of economies experiencing declines in per capita output since 1870' (The World Bank, 2020). The Bank repeated the crisis messaging of the WHO and UN at the time, arguing for 'the urgent need for health and economic policy action, including global cooperation, to cushion its consequences, protect vulnerable populations, and strengthen countries' capacities to prevent and deal with similar events in the future' (The World Bank, 2020).

In October 2020, eight months into the pandemic, three UN agencies (International Labour Organization, Food and Agriculture Organization, and

International Fund for Agricultural Development) and the WHO released a joint statement declaring:

> The COVID-19 pandemic has led to a dramatic loss of human life worldwide and presents an unprecedented challenge to public health, food systems and the world of work. The economic and social disruption caused by the pandemic is devastating: tens of millions are at risk of falling into extreme poverty, while the number of undernourished people, currently estimated at nearly 690 million, could increase by up to 132 million by the end of the year. (World Health Organization, 2020h)

This statement, like the UN Secretary-General's call earlier that year, conveyed a sense of urgency and panic. It noted: 'Millions of enterprises face an existential threat', and 'Nearly half of the world's 3.3 billion global workforce are at risk of losing their livelihoods', and that 'The pandemic has been affecting the entire food system and has laid bare its fragility' (World Health Organization, 2020h). The organizations then offered a plea for 'global solidarity and support, especially with the most vulnerable in our societies, particularly in the emerging and developing world', noting that 'Only together can we overcome the intertwined health and social and economic impacts of the pandemic and prevent its escalation into a protracted humanitarian and food security catastrophe, with the potential loss of already developing gains' (World Health Organization, 2020h). Finally, they stated, 'We must recognize the opportunity to build back better', and that 'We are committed to pooling our expertise and experience to support countries in their crisis response measures and efforts to achieve the Sustainable Development Goals' (World Health Organization, 2020h).

Secondly, in line with the UN Secretary-General's call, many governments *did* implement stimulus measures, including direct payments to individuals (in the US, called economic impact payments or stimulus checks or payments), payments to business to retain workers (in Australia, called Job Keeper Payments), increased unemployment benefits, 'forgivable loans' to small businesses and/or corporations and state and local governments, and short-term work programmes to preserve jobs and support demand (as in Germany) (Jordà and Nechio, 2023). These measures no doubt prevented many people falling into poverty (at least to begin with), kept some businesses afloat during this period, and allowed many people to keep their jobs. They were arguably critical in averting a legitimation crisis, which is inherent in capitalist societies, involving a widespread loss of citizens' confidence in administrative institutions, which had the potential to undermine the 'mass loyalty' or public support that underpins rule in liberal democratic capitalist societies (Habermas, 1976).

Yet, at the national level, support packages were selective in their application and uneven in their impacts, so that some groups, such as casual or gig workers, including those employed in the creative economy such as visual artists, musicians, and other performers, were often left to fend for themselves, while those remaining in employment or who were furloughed and maintained a link with their employer fared relatively well. (See, for example, Despard et al, 2021 for the US.) Platform workers were strongly impacted by the COVID-19 pandemic, with one third of women (32 per cent) and two-fifths of men (39 per cent) losing their paid jobs during the initial lockdowns, and many having to leave their accommodation as a result (European Institute for Gender Equality, 2021: 6). The selection of businesses and individuals for support revealed much about the priorities and values of policy makers in the different countries.

As with past economic crises, women not only suffered more job losses overall (especially those with less education) but were confronted with increased housework and childcare responsibilities following the rapid switch to remote working and home schooling (Del Boca et al, 2020; Dunatchik et al, 2021). According to one estimate, women were 1.8 times more vulnerable to job losses than men (Madgavkar et al, 2020). Women also experienced increased incidences of domestic violence (World Wide Web Foundation, 2020). The responses, which were often perceived as discriminatory or shown to have discriminatory effects, may have fuelled resentment among those who were made redundant or lost their businesses, or who otherwise were detrimentally impacted, potentially contributing to acts of lockdown resistance, vaccine resistance, and other demonstrations of civil unrest. Some COVID-19 policies, including restrictions on gatherings, international travel, and internal movement, and workplace closures, have been shown to be linked to increased numbers of protests and riots during the pandemic (Wood et al, 2022). Indicators of grievance and acts of dissent and resistance were also apparent during the SARS and Ebola outbreaks in West Africa and Canada (van der Zwet et al, 2022).

Three years after the declaration of the emergency, the stimulatory measures were judged by some researchers and economists to have been excessive. For example, in Australia, in April 2023, research undertaken by the Australian National University lent weight to the argument that 'Pandemic-era government stimulus measures such as JobKeeper dramatically overcompensated for lost incomes and, when combined with rock-bottom interest rates, pushed inflation 3 percentage points higher than it needed to be' (Read, 2023). In the same month in the US, Agustin Carstens, general manager of the Bank for International Settlements, observed that the rise in inflation over the previous two years had been 'large, sudden and global' and 'at levels unseen for generations' (Carstens, 2023). Moreover, Carstens said, financial systems were under strain, with recent bank failures the most obvious example. High inflation and financial stress were 'emerging in tandem'. In his view this was a symptom of 'a broader and deeper problem'; namely, that 'monetary and fiscal policy have tested the boundaries of what I call the "region of stability"', or the management of the balance between these policies needed for macroeconomic and financial stability (Carstens, 2023: 1).

Carstens went on to note that governments' stimulus measures, with interest rates held at zero and often below, had perverse effects.

> Fiscal stimulus since the start of the pandemic has exceeded 10% of GDP in many advanced economies – a push previously seen only in wartime. In many countries, households received large transfers, which in some cases exceeded the income hit from the pandemic. Firms benefited from generous subsidies to keep unutilised workers on the payroll, their debts were guaranteed. Because these measures added to demand at a time when supply was artificially constrained, they boosted inflation but did little for growth. The stimulus continued, and in some cases grew, long after economies started to recover. (Carstens, 2023: 2)

Central banks' efforts to 'tame inflation', with multiple increases in interest rates, meant that many businesses struggled to survive and financial institutions, such as Credit Suisse in Switzerland, and Silicon Valley Bank and Signature Bank in the US, teetered on the verge of collapse. It could be argued that the pandemic merely accentuated systemic weaknesses in economies rather than having a direct causal effect. But there is little doubt that pandemic measures have had reverberating implications, especially with supply chain disruptions and labour shortages, which fueled inflation. The timing of the war in Ukraine and related energy supply shortages can be seen as being contributed to by COVID-19, in that the economic and public health crisis had weakened all states, except energy-producing and -exporting countries, which provided a 'window of opportunity' for the Kremlin's 'special operation' (Fontanel, 2023).

Thirdly, at the global level, the ability of countries to implement stimulus measures varied greatly. While some countries were able to draw on reserves or borrow funds that supported businesses, workers, and households, others struggled or provided little or no support. The UN looked to the G20 to help address the pandemic crisis, but the policies its members adopted mostly reflected their own interests. The G20, to whom the UN Secretary-General's appeal was made, is an intergovernmental forum comprising the world's largest economies, which, as noted by the UN Secretary-General, accounts for a large proportion of the world's gross national product. In early 2023, the G20 comprised 19 countries, namely, Argentina, Australia,

Brazil, Canada, China, France, Germany, India, Indonesia, Italy, Japan, Republic of Korea, Mexico, Russia, Saudi Arabia, South Africa, Turkey, United Kingdom and United States, and the European Union (Department of Foreign Affairs and Trade, 2023). When G20 countries began to raise interest rates in 2022, this disproportionally negatively affected poorer countries. As Triggs (2021) argues, the G20's policies had contributed to creating a 'two-speed global economy', with poorer countries then at risk of suffering the financial turbulence created by the decision of 'Many rich world central banks tapering quantitative easing programs and raising interest rates' (Triggs, 2021).

The G20 as a crisis mechanism

If one is to understand the nature of the G20's response to COVID-19, I believe it is important to appreciate the organization's history and operations. G20 itself is a *product of crises*, being created in the late 1990s in response to a series of debt crises beginning with the Asian financial crisis in 1997, which had cascading impacts on other economies (Ibbitson and Perkins, 2010). As the G20's web page explains, the group was founded in 1999 after the Asian financial crisis 'as a forum for the Finance Ministers and Central Bank Governors to discuss global economic and financial issues'. It was then upgraded to 'the level of Heads of State/Government in the wake of the economic and financial crisis of 2007, and, in 2009, was designated the "premier forum for international economic cooperation"' (G20, 2024).

While the G20 'initially focused largely on broad macroeconomic issues ... it has since expanded its agenda to inter-alia include trade, sustainable development, health, agriculture, energy, environment, climate change, and anti-corruption' (G20, 2024). From this list it can be seen that the G20, which comprises the most powerful nations in the world, has a very broad agenda, bringing together non-governmental organizations, think-tanks, women, youth, labour, business, and researchers. As a crisis response mechanism, the G20 draws on the considerable economic and political power of these nations to stabilize the global system – which ultimately serves their own interests.

Compared with formal international organizations, the G20 operates as an informal 'club' which is not governed by an international treaty or fixed constitution and has been criticised for its under-representation of developing countries and lacking legitimacy (Gao and Wouters, 2023: 78). This should be kept in mind when assessing the international reaction to the COVID-19 pandemic. The 'club'-like character of the G20, it has been argued, has led to it focusing too much on developed-world problems and failing to respond to developing world issues. In December 2021, nearly two years into the pandemic, one commentator accused the group of recycling the G7's agenda (this group comprising the US, Canada, Japan, the UK, France, Germany, and Italy) (Triggs, 2021).

The G20 was also criticized for not dealing with the challenges facing Asia's developing economies, in that 'Rich countries are expected to reach their pre-pandemic levels of output by 2024, while poor countries will remain 5.5 per cent behind their pre-pandemic levels' (Triggs, 2021). Weak healthcare systems, combined with weak social safety nets and limited fiscal and monetary policy, meant that developing countries were bound to fare poorly from COVID-19. Developed countries did relatively little to support these countries, with most of the former's financial supports, like currency swap lines, going to other rich countries. The rollout of vaccines in 2021 mostly benefitted the high- and middle-income countries, notably members of the G20, which engaged in 'vaccine nationalism' entailing the signing of agreements with pharmaceutical companies to meet their own citizens' needs. As the Triggs report noted in 2021, 'Almost 60 percent of the population in advanced economies are fully vaccinated', with citizens receiving their booster shots whereas 'In poor countries more than 95 percent of the population remain unvaccinated' (Triggs, 2021).

The rollout of COVID-19 vaccines in 2021 exposed the gulf between nations in their levels of wealth and power. Studies undertaken by Duke University, published in April 2021, a year into the pandemic, found that most COVID-19 vaccines had been administered by a small number of high- and middle-income nations and regions, including the United States (US), United Kingdom (UK), the European Union (EU), China, and India. The report noted, 'Just four nations or regions with less than half the world's population have administered seventy percent of all COVID-19 vaccine doses, while the poorest countries have barely begun vaccinating due to lack of funding and supply. The wealthiest nations have locked up much of the near-term supply' (McClellan et al, 2021).

The extent of vaccine inequality highlighted by this research is startling, but not surprising, given global disparities in wealth and technological capacity. As indicated in the report, various issues contribute to this inequality, but perhaps most important of all is the intellectual property protections for patents that are owned by companies that are mostly located in richer parts of the world, especially the US. To increase global access to COVID-19 vaccines and therapies during the emergency, India, South Africa, and some other nations called for the temporary waiving of TRIPS (Trade-Related Aspects of Intellectual Property Rights) covering patents, industrial designs, trade secrets, and regulation data for COVID-19 vaccines and therapies for the duration of the pandemic (World Trade Organization, 2020).

However, despite these calls and support from a group of eminent former heads of state and scientists who made appeals to US President Biden, pharmaceutical and biotech companies resisted a patent waiver or sharing their expertise. These companies, most of which are based in resource-rich countries, had invested heavily in vaccine development, which is protected through patents, and were reluctant to loosen patent and intellectual property protections on coronavirus vaccines, which would affect their future profits and undermine their business model (Stolberg et al, 2021). The issue of vaccine nationalism, whereby countries signed agreements with pharmaceutical companies to meet the needs of citizens in their own countries first, highlighted the limits of the notion of global citizenship called for by the UN in responses to a global pandemic. It underlined the tendency to view vaccines and essential medicines as a market commodity rather than as a public good, which ultimately reinforced inequities in access and disparities in health and wellbeing (Katz et al, 2021).

Critical evaluations of pandemic preparedness and responses

The issue of vaccine nationalism was flagged in a report of an Independent Panel on pandemic responses, commissioned by the Director-General of WHO in May 2020. This Panel, which was led by Her Excellency Ellen Johnson Sirleaf and Helen Clarke, revealed many shortcomings in pandemic preparedness and responses, especially during the pandemic's early months. As outlined in its initial report, from mid-September 2020 the Panel began to 'review the extensive literature, conducted original research, heard from experts in 15 round-table discussions and in interviews, received the testimony of people working on the front lines of the pandemic in town-hall-style meetings, and welcomed many submissions from its open invitation to contribute' (The Independent Panel for Pandemic Preparedness and Response, 2021: 8). The Panel also produced a companion report outlining 'thirteen defining moments which have been pivotal in shaping the course of the pandemic', as well as a series of background papers based on its research which included 'a chronology of the early response' (2021: 8).

The Panel noted that, as of 28 April 2021, '148 million people were confirmed infected and 3 million have died in 223 countries, territories and areas', '90% of schoolchildren were unable to attend school', '10 million more girls were at risk of early marriage because of the pandemic', 'gender-based violence support services have seen five-fold increases in demand','115–125 million people had been pushed into extreme poverty', and 'US$10 trillion output is expected to be lost by the end of 2021, and US$22 trillion in the period 2020–2025, the deepest shock to the global economy since the Second World War and the largest simultaneous contraction of national economies since the Great Depression of 1930–1932' (The Independent Panel for Pandemic Preparedness and Response, 2021: 10).

The Independent Panel documented failures in pandemic preparedness, including 'insufficient fundings at national, regional and global levels before the pandemic, and the *slow flow* of funding for response once the PHEIC [Public Health Emergency of International Concern] was declared' (2021: 56; bold in original). But, it said, national responses varied greatly, with countries that were best able to manage the disease having sought scientific guidance, engaged with community workers and community leaders and vulnerable and marginalized communities, and adopted more socially inclusive, 'whole-of-government and whole-of-society approaches' (The Independent Panel for Pandemic Preparedness and Response, 2021: 58). In its assessment, 'This pandemic has shaken some of the standard assumptions that a country's wealth will secure its health' and that 'Leadership and competence have counted more than cash in pandemic responses' (2021: 11). In its view, 'Many of the best examples of decisive leadership have come from governments and communities in more resource-constrained settings' (2021: 11).

The Independent Panel noted that inequity in access to vaccine and vaccine nationalism was 'one of today's pre-eminent global challenges' (2021: 12). High-income countries, including Australia, Canada, the US, New Zealand, and members of the EU 'have been able to secure vaccine doses that would be enough to cover 200% of their populations' through 'bilateral deals with manufacturers to secure existing and future stocks', whereas in the poorer countries (at the time of finalizing the report) 'fewer than 1% of people have had a single dose of vaccine' (2021: 12, 41).

Six months after the publication of this report, in November 2021, the Independent Panel released *Losing Time: End this Pandemic and Secure the Future*, which documented progress in the interim period but noted that it was 'not fast or cohesive enough to bring this pandemic to an end across the globe in the near term, or to prevent another' (Sirleaf and Clark, 2021: 1). The report again documented the issue of vaccine inequality, commenting that the 'immunization gulf between the richest and poorest countries of the world jeopardises the health of everyone on the planet'. It estimated that 'More than 67% of the population of all high-income countries has been fully vaccinated against COVID-19, but in low–income countries fewer than 5% have received even one dose, and that figure hovers even lower in many' (2021: 9). This level, it said, was well below the WHO target of 40 per cent of the population of each country to be fully vaccinated by the end of 2021 and 70 per cent by mid-2022. This was deemed to be a 'minimum achievable goal based on vaccine supply forecasts', but the rates needed to be higher to protect the overloading of healthcare systems. While, as the report stated, by 1 September of 2021 there had been some redistribution of vaccine doses, this did not meet the WHO's target. Moreover, the capacity of low- and middle-income countries to purchase vaccines had been restricted by 'confidential high-cost deals between manufacturers and wealthy countries as they add booster doses to their immunization programmes, despite powerful arguments against this on equity grounds' (2021: 9).

What was evident at this time (and discussed in this report) was that short-term national interests and the interests of the pharmaceutical industry who had put profit before public health, and bought patents and developed them, worked to the detriment of lower- and middle-income countries. Events also revealed that the WHO had limited powers and lacked the independence to effectively implement its recommendations. One of the proposals at this time was to develop an international treaty or treaties to address pandemic preparedness and responses which were found to be wanting, along with improved financing of the WHO (Sirleaf and Clarke, 2021: 23–24). This report concluded with yet another call for concerted global action and 'multilateral leadership', including an 'Independent Panel for Pandemic Preparedness and Response' to provide a starting point for 'urgent reforms to strengthen the global health architecture' (2021: 26).

Increase in wealth disparities

Crises are known to produce economic, health, and social inequalities. The global financial crisis beginning in 2007 disproportionately affected lower-income, less educated, and minority households (Pfeffer et al, 2013). The pandemic crises of SARS (2003), MERS (2012), Ebola (2014), Zika (2016), and the Influenza pandemic of 1918 generated economic, gender, age, health, and/or other inequalities (Roberts and Tehrani, 2020; Brzezinski, 2021). In the case of COVID-19, by early 2023, accumulating evidence was showing a massive increase in wealth and health disparities directly or indirectly linked to crisis measures.

A report published by Oxfam in January 2023 noted that 'Since 2020, the richest 1% have captured almost two-thirds of all new wealth – nearly twice as much money as the bottom 99% of the world's population' (Oxfam, 2023: 7). As the world confronted an 'unprecedented "polycrisis"' – including rising costs of living, climate breakdown and continuing impacts of COVID-19 – 'poverty has increased for the first time in 25 years' (2023: 7). Some months earlier, the World Bank also released a report which announced among its many findings that 'The COVID-19 pandemic dealt the biggest setback to global poverty in decades', with extreme poverty increasing from 8.4 per cent in 2019 to 9.3 per cent in 2020, or more than 70 million people (World Bank Group, 2022: xxi).

According to Credit Suisse Research Institute's *Global Wealth Report* (2022), a growth in global wealth by the end of 2021 (9.8 per cent over 2020) and rising inequality in 2020 and 2021 'is probably due to the surge in the value of financial assets during the COVID-19 pandemic' (Shorrocks et al, 2022: 5, 7). As the authors of this report explain:

Since the start of 2020, the evolution of household wealth has been driven by the economic repercussions of the COVID-19 pandemic and the actions taken by national and international agencies to mitigate the negative impacts on businesses and individuals. The first few months of 2020 followed a familiar script. Expectations of sharp falls in economic activity and trade caused share prices to dive and global household wealth to fall by 4.4% between January and March. However, robust and prompt (often pre-emptive) responses by governments and central banks helped to stabilize financial markets, enabling the earlier household wealth losses to be largely reversed by mid-year. The generous financial support given to households in many advanced countries, coupled with lower interest rates and limitations on consumption opportunities, boosted household savings and laid the foundations for share price and house price increases, which resulted in significant rises in household wealth throughout the world. This good fortune for wealth holders contrasted sharply with the medical challenges and the broader economic hardships that prevailed in most of the world. (Credit Suisse Research Institute, 2022: 7)

In the World Bank's assessment, '2020 marked a historic turning point – an era of global income convergence gave way to global divergence', and incomes in the poorest countries fell more than in those in the rich countries, the result being that 'global inequality rose for the first time in decades' (World Bank Group, 2022: xxi). It argued that measures such as taxes, transfers, and subsidies tend to be less successful in benefitting the poor than is the case in high-income countries. The reason for this, according to the World Bank Group, is that in two-thirds of these countries the income in poor households tends to fall by the time taxes have been paid and transfers and subsidies are received, which is a consequence of the relatively large informal sector in those economies and reliance on indirect taxation (for example, sales and excise tax), and 'income transfers are too low to compensate' (World Bank Group, 2022: xxii).

The World Bank Group report noted that extreme poor households had less access to electricity and mobile phones on the eve of the COVID-19 pandemic, which meant that they were disadvantaged in access to online content and services which became critical for school and work as people around the world became highly dependent on the internet (2022: 39–40). Hence, children who were unable to attend school due to lockdown restrictions missed out on education during a critical phase of their development, which is likely to further reinforce inequalities. (See also Schleicher, 2020.) The report suggests that the non-monetary dimensions of the pandemic and its impacts (of which gaps in education are an important aspect) 'may ultimately prove to be more costly than the monetary dimensions' (World Bank Report, 2022: 6).

Substantial increases in the cost of living, including housing, in many rich countries during and in the immediate aftermath of COVID-19, much directly or indirectly linked to pandemic measures, including stimulus packages, have reinforced inequalities. Lockdowns and other restrictions on social movement during the first year of COVID-19 resulted in many people losing their jobs or businesses, which meant that they could no longer pay their rents or mortgages or meet their utility payments. Many turned to social services and community organizations for support. In the US, a 2021 survey undertaken by the Social Policy Institute at Washington University uncovered the disproportionate impact of the pandemic on renters, parents, the unemployed, and those from racial/ethnic minority groups (Social Policy Institute, 2021). The survey, which was released in September 2021, found that in those households where someone had lost a job, financial hardship was three times higher than among those who had not. It also found that Black and Hispanic households were 10 per cent more likely to experience the burdens of housing cost (Social Policy Institute, 2021). The trends identified in this survey broadly mirrored those reported in a number of other countries. Housing costs surged in many countries between 2020 and 2021, which has been attributed to cheap interest rates during this period. This led some people to purchase properties that they would not otherwise have been able to afford and put many people at risk of future rises in interest rates, which began to be implemented in some countries in 2022 and 2023 in response to the aforementioned stimulus measures.

Those who fell into poverty or whose poverty was exacerbated during COVID-19 have been vulnerable to exploitation in the informal economy, or so-called shadow economy, which is not taxed or monitored by governments. Because the shadow economy operates outside the law in non-public domains, its dimensions are difficult to ascertain. One part of this economy is human trafficking and slavery. A report, *Global Slavery Index*, published by Walk Free in May 2023, comments:

> Mass unemployment, high personal debt, and limited government support created opportunities for traffickers to prey on growing numbers of people who have been pushed into survival mode during the COVID-19 pandemic. Many of those desperate to stay afloat have been driven into forced labour and bonded labour, often in sex work and brick kilns. With many industries yet to return to pre-pandemic operations, pathways out of modern slavery remain limited. (Walk Free, 2023: 19)

According to the authors of this report, for many, increased time online during the global lockdowns was capitalized on by those with criminal intentions who, shielded by online anonymity, were able to use 'deceptive job advertisements and targeted outreach activities using chat features to recruit vulnerable people into exploitative labour'. Adolescent girls, it is noted, were especially targeted and 'lured into forms of sexual exploitation under the guise of building a romantic relationship'. As the report concluded, the pandemic was 'a perfect example of how crises compound other crises', with the effects falling disproportionately on the most vulnerable and exacerbating existing inequalities that underpin modern slavery (2023: 19).

While many people became poor or poorer during the period of the declared emergency, some people increased their wealth or became extremely rich. This is shown in patterns of spending on expensive houses, cars, jewellery, boats and holidays, and other luxury items and services, which returned to or surpassed 2019 levels, following a market contraction in 2020 – despite recessionary conditions (D'Arpizio et al, 2023). One of the most evident indicators of increased extreme wealth post-COVID-19 was the pattern of private/charter jet usage. The surge in travel via private jets began in 2022 when many countries were emerging from what was widely seen as the worst of the pandemic. In April 2023, an Australian news article noted that the chartered plane market 'is booming post-Covid, offering speed and convenience to a privileged few – and generating tonnes of CO2' (Belot, 2023). The article draws on data from a federal government report that showed that 'fixed-wing charter operators carried 319,758 passengers in January 2023, compared to 239,044 in January 2022 and 249,430 in pre-COVID January 2020' (that is, a 28 per cent increase) (Australian Government, 2023: 1). The article said that 'aviation experts' attributed the recent increase to 'wealthy people chartering private planes' who were 'unwilling to risk cancellations and delays' (amid a reported increase in delayed commercial flights and declines in service in the wake of the pandemic) and were using the jets 'for holidays, interstate meetings, and relocating their pets in luxury' (Belot, 2023). A pilot from one private operator in Melbourne was cited as saying that business is booming since the pandemic 'thanks to "time-poor, wealthy individuals who are not price sensitive"' (Belot, 2023).

Australia's experience of private jet use post-COVID-19 mirrors that in other countries. For example, in the UK, the number of private jets departing from the UK 'increased by 75% between 2021 and 2022 to 90,256 flights, emitting 500,000 tonnes of CO2', according to a Dutch environmental consultancy group (Horton, 2023). The article reporting this finding referred to a Transport and Environment (2021) study showing that 'private jets are five to 14 times more polluting than commercial planes per passenger, and 50 times more polluting than trains'. It also referred to earlier research which found that '50% of aviation emissions were caused by 1% of the world's population' (Horton, 2023). In January 2023, a market report said that 2022 was 'the strongest year for private jet sales in the history of the International Aircraft Dealers Association' – although it was noted that 'market conditions are cooling' and that 'buyers are being more cautious', with concerns about an imminent recession (Kelly, 2023). While private jets benefit only a relatively small number of wealthy citizens, the fact that they have increased their use of this highly polluting and environmentally harmful technology conveys much about the extremes of inequality and wealth following the declaration of the pandemic.

Conclusion

I have referred to some of the accumulating evidence on the inequalities and disparities directly or indirectly linked to the COVID-19 crisis during its first two years. These inequalities arose directly from the initial measures to mitigate the risks, such as lockdowns and border closures, which led to job losses and business failures, and then from the fiscal stimulus that fuelled inflation and created a 'two-speed global economy'. Clearly, some groups have benefitted greatly from crisis conditions, to begin with at least, as seen in evidence on the growth in extreme wealth during the first two years following the pandemic's onset, while others fell into poverty or became poorer.

It will be some time before the inequitable effects of the pandemic measures become fully apparent, and they may then be difficult to evaluate, given the complexities of teasing apart the contributors to inequality. The ongoing impacts of the crisis on those who lost their jobs (including many gig workers, musicians, artists, and others previously employed), many small business owners, and children whose educations were disrupted due to home schooling, are incalculable. Yet national lockdowns also provided ideal conditions for the rapid advance of AI-based technologies, which has both bolstered the profits of Big Tech and smaller companies that utilize digital platforms and created new options for work, schooling, housing, and recreation. In the next chapter, I explore how the pandemic crisis contributed to technological advance and some of the consequences that have followed in its wake.

5

Pandemic Crisis and Technological Change

Crises, regardless of their origins, often serve as catalysts for technological development. This has been evident with the COVID-19 crisis, which led to huge investment in and research on new vaccines, diagnostic tests, and technologies of tracking and tracing to help prevent the spread of disease. Crucially, the crisis massively boosted investment in digital technologies which became essential tools not just for mitigating the risks of infection and disease but also for undertaking work, education, and recreational pursuits during the lockdowns. However, while the advances have been widely lauded for the benefits they delivered during the pandemic – especially in preventing illness and saving jobs and lives – they have impacted upon people's lives to varying degrees and created new personal and social risks.

In this chapter, I examine the dimensions and socio-political implications of these technological changes, focusing on the three-year period following the onset of COVID-19. I discuss how the crisis accelerated the collection of data that enabled the rapid advance of artificial intelligence (AI), which has potentially profound consequences for societies. As I explain, these advances have been underpinned by three interlinked processes: the rapid collection of data for the purpose of pandemic surveillance; the sudden surge in the population of users of digital technologies; and massive investment by Big Tech companies seeking to gain a competitive 'edge' in the 'race' to exploit the opportunities provided by the crisis. I explore how these processes played out in the aftermath of the declaration of the pandemic, especially its early phases when lockdowns were implemented in many countries, highlighting the implications not just for the character of the innovations but for how citizens conduct their everyday lives.

Digitization during COVID-19

As I noted in Chapter 1, COVID-19 was the first pandemic to occur during the digital age, and this has profoundly shaped people's experiences of and official responses to the pandemic. Digital technologies are powered by data whose value escalated as scientists sought to develop powerful models to predict risk, and the potential of AI soon became apparent. As surveillance studies sociologist David Lyon (2022) explains, the pandemic arrived at a time when big data was already prized for its apparent value in solving problems in many areas of life, including in government. This data provided the raw material not just for public health surveillance, namely disease prevention and control, but also for political surveillance, including making visible and monitoring and policing specific groups or whole populations (2022: 24–32).

However, during the pandemic, the distinction between these different forms of surveillance was often blurred as authorities expanded their powers and citizens became reliant on mobile phones and apps for contact tracing, QR (quick response) check-ins, and updates on the pandemic. Digital innovations enabled researchers to track the dynamics of social behaviour via analysis of citizens' online data trails or digital footprints. This provided authorities with insight into how individuals and communities were responding to outbreaks and measures, enabling governments (for example, China, Israel, Hong Kong, and South Korea) to adopt active surveillance measures including the use of personal data and the downloading of data via smartphone apps to track people's movements and contacts (Sweney, 2020).

The advance of data-sharing systems

In March 2020, the same month that the WHO had declared the pandemic, it was reported that the mobile phone industry had explored 'a global data-sharing system that could track individuals around the world, as part of an effort to curb the spread of Covid-19' (Kirchgaessner, 2020). The plan, considered by GSMA (Global System for Mobile Communications), an organization that represents the interests of and sets international standards for 750 mobile phone operators, was reported to enable authorities 'to monitor movements and potentially track the spread of the disease across borders' (Kirchgaessner, 2020). The UK Government was said to be 'in talks with UK mobile phone companies to potentially use anonymous location and usage data to create movement maps, with a 12- to 24-hour delay, to discover whether the public are abiding by lockdown rules' (Sweney, 2020). Vodafone quickly responded to the call, collaborating with national governments and supranational entities to deliver new metrics for the pandemic which, one of its researchers notes, has 'proven to be useful for a number of external entities' (Lourenco et al, 2021).

An Organisation for Economic Co-operation and Development (OECD) report, published in April 2020, elaborated on the proposal:

As COVID-19 continues to take human lives and jolt the global economy, governments are urgently seeking innovative new tools to inform policy and tackle the crisis. Digital solutions based on geolocation data are emerging to help authorities monitor and contain the spread of the virus. Some are fed by mobile call data records (CDRs), ie data produced by telecommunication service providers on telephone calls or other telecommunications transactions, which provide valuable insights into population movements. As network operators serve substantial portions of the population across entire nations, the movements of millions of people at the spatial and temporal scales can be measured in near real-time. The resulting information and trends are invaluable for governments seeking to track the COVID-19 outbreak, warn vulnerable communities, and understand the impact of policies such as social distancing and confinement. (OECD, 2020a)

In 2020, a number of OECD countries began to share CDR-based geolocation data with governments 'in an aggregated, anonymised format', citing various examples of where this had occurred. It also announced the launch of new mobile applications of COVID-19 tracking which, it noted, were increasingly being developed as open source, being produced by partnerships involving tech companies, universities, clinicians, and public authorities (OECD, 2020a). Finally, it referred to facial recognition being 'one of the most frequently used biometrics in a number of countries to monitor the spread of COVID-19', which has 'enabl[ed] authorities to reduce the use of identification technologies that require physical contact (such as iris scans and fingerprints)'. Facial recognition, it noted, 'can also be paired with other technologies, including thermal imaging enhanced by artificial intelligence, to better track citizens who may test positive for COVID-19' (OECD, 2020a).

While the title and content of this report highlight concerns with protecting privacy, and the need for 'privacy-by-design' to address risks, this could not be guaranteed in the state of exception which accorded authorities extraordinary powers and suspended citizens' rights. Indeed, crisis conditions enabled *unparallelled access* to citizens' data, which was needed to 'power' the facial recognition and other AI-enabled technologies used during COVID-19. Researchers were able to track citizens' 'digital footprints' and telecommunication companies, such as Vodafone, shared geolocation data, allegedly to provide insight into the level of population mobility and assess the impact and effectiveness of containment measures (Lourenco et al, 2021). (See also Chapter 3.)

The WHO is also keen to advance data-sharing to assist with pandemic efforts. Its vision was articulated in a White Paper published in 2022 as part of a series of proposals for 'Strengthening the global architecture for health emergency preparedness, response and resilience' (World Health Organization, 2022a). In his foreword to the White Paper, WHO Director-General Tedros Adhanom Ghebreyesus notes that the COVID-19 pandemic has laid bare the 'deep flaws in the world's defences against health emergencies', adding that 'The world was, and remains, unprepared for large-scale health emergencies' (World Health Organization, 2022a: iv). The White Paper's recommendations include the strengthening of data-sharing platforms that will connect and integrate diverse sources of information, and to create advanced analytical and modelling tools (2022a: 21). Notably, the White Paper flags 'infodemic management' as critical in the effort to inform communities and build trust for 'countermeasures such as vaccines' (2022a: 7–8, 18, 22).

However, this enthusiasm for data sharing during COVID-19 also created new security risks. In the first year of COVID-19, various reports documented increases in cybercrime and other online harms and abuses (especially to women and children), and the generation of rumours and fake news facilitated by AI (for example, Interpol, 2020; Parliament of Australia, 2020). There was also a series of highly publicized cyberattacks – in Australia, notably those involving Optus, Medibank, and Woolworths in 2022, when personal information from millions of citizens was leaked (Office of the Australian Information Commissioner, 2023). Digital innovations developed at such a rapid pace during the pandemic that policy makers and regulators struggled to ensure the security of citizens' personal data.

A security specialist, Kris Lovejoy, who works for Kyndryl, a major global provider of information technology (IT) infrastructure services, is reported to have found in her work at Ernst & Young that '60 percent of the world's firms have introduced new technology to enable COVID-19 working from home and new ways of working with clients'; however, 'Of that, 60 per cent either abbreviated or altogether skipped over the implementation of security control in and around that technology' (Burton, 2023). According to Lovejoy, 'attackers had shifted their attention to large supply chain platforms' (Burton, 2023). As noted by Lovejoy:

> They are attacking the technology providers themselves and looking for the ways in which they could integrate themselves into the technologies that are being built and delivered to consumers of that technology. We don't realise that those technologies now have backdoors or inherent weaknesses that were implemented and deployed by the attacker so that they would be able to get access into those programs. That's why I think there's a lot of concern today about the supply chain and the resilience of the supply chain, particularly when it comes to technology vendors. There is a sense that a lot of the technology that was built and introduced into organisations who hadn't checked to see whether or

not what they were buying had the appropriate integrity and security controls. (Burton, 2023)

In Lovejoy's view, while many enterprise platforms had good security during the COVID-19-induced rapid shift to digital technology, some of the connected apps did not, especially internet-of-things devices such as beacons and cameras. Lovejoy also expressed concerns about file-sharing sites, many of which had been created during the pandemic to enable collaboration 'without appropriate authentication requirements in and around those sites', such as backup storage and archive data that often includes a lot of personally identifiable information (Burton, 2023).

As often happens with new technologies, innovations are introduced before security safeguards are in place. The rush to adopt new data-sharing systems during the pandemic meant that some organizations were taking shortcuts and not ensuring that the technologies they were introducing did not present risks to their users. However, the Big Tech companies themselves arguably contributed to the problem. These companies are often described as being in a technological 'arms race' which greatly intensified during the pandemic and increased the pressure to bring innovations to the market. The use of the 'arms race' metaphor conveys the issues at stake and the 'winner-takes-all' conception of technological advancement. The 'reward' in this case is legal monopoly control of software – so-called proprietary (or non-free or closed source) software – as opposed to open source software, where everyone is free to use, copy, or change and, it is argued, hopefully, improve it (Bessen, 2022).

According to Bessen, this investment in proprietary software enables companies to 'compete by leveraging complexity'; that is, by offering a greater variety, and more product versions and features, and thereby to differentiate themselves from competitors, which substantially alters the nature of markets and the structure of industry. Big companies like Meta/Facebook, Amazon, Apple, Netflix, and Google are best placed to develop proprietary software which has reduced competition, slowed the diffusion of technical knowledge, and led to market domination by a few 'superstar' firms (Bessen, 2022).

The technological 'arms race' has had negative consequences for AI research itself. As noted by the UK House of Commons in its report on reproducibility and research integrity, published in April 2023, competition in AI research and development has pernicious effects, including the pressure in the field of AI to publish more and at a faster pace than in other fields, with the result that there is a greater risk of 'research garbage' because of the rapid development of the field (House of Commons Science, Innovation and Technology Committee, 2023).

The House of Commons report received evidence indicating 'an array of reproducibility issues with the methodologies employed in the artificial intelligence (AI) field, especially around code', which, it was noted, 'was in line with published findings on the artificial intelligence research system – the 2021 "State of AI report" [which] found that only 26% of AI papers that year had made their code available, which has concerning implications for transparency' (House of Commons Science, Innovation and Technology Committee, 2023: 16–17). The report noted that 'This is in part exacerbated by especially negative incentives in the AI field, which are more pronounced than in other disciplines' (2023: 17). Yet, notwithstanding these concerns, AI has proceeded at a rapid pace, being enabled by changes in work and education which followed from pandemic crisis measures.

Technology's impacts on everyday life

It may be some time before it is known whether all, some, or many of the technological changes occurring during the pandemic become 'normalized'. Following the lockdowns, most students returned to their schools or universities and many businesses encouraged or insisted that their staff return to their physical places of work. However, a global survey of business executives undertaken in late 2022 revealed that many businesses were then settling for hybrid work models, where staff would be required to be in the office for some days during the week (The Conference Board, 2023: 5–6). The work-from-home/hybrid work picture is complicated by the fact that the character and extent of changes varies between different areas of work and different countries (Vyas, 2022). As noted, many companies embraced or increased their use of digital tools of all kinds during the pandemic to establish new ways of doing business, which were not so strongly tied to a central office or factory. Schools and universities employed Zoom, Microsoft Teams, and other platforms to conduct classes, while casual and contract workers adapted as best as they could, sometimes launching new business ventures and/or experimenting with new ways of working or practising their craft.

While the shift towards home-based or hybrid/flexible working was underway in many countries well before COVID-19, the trend accelerated with the onset of pandemic-induced lockdowns and other restrictions. Between mid-2019 and mid-2020, international internet traffic increased 48 per cent, which is largely attributed to the pandemic (Altman and Bastian, 2021). The sudden reliance on digital technologies enabled some companies to increase their efficiencies and profits (through time-space compression (Harvey, 1990)), and some individuals to survive and sometimes thrive; but others who were unable or too slow to adapt (for example, without an existing online platform) have been disadvantaged or excluded from business or access to educational resources or options for sociality. In the case of businesses, three years after the onset of the pandemic, some found that they could operate with fewer staff and less office or retail space. This has impacted upon some parts of the commercial property sector, especially in areas of employment where staff were previously expected to be in the office most days of the working week (Ezrati, 2023).

The pandemic provided the opportunity for many companies to trial new, more flexible, sometimes more profitable ways of working; for example, using online sales rather than physical stores. People around the world became, in effect, subjects of a natural experiment (see Chapter 6), with the disruptive effects of the crisis leading many to question taken-for-granted aspects of their lives. Some took the opportunity to relocate, which enabled them to fulfil their preferred 'work–life balance' (the so-called tree/sea change phenomenon) if they had the resources to do so.

When COVID-19 arrived, some people in some countries decided to abandon cities and move to outer suburban and regional areas, for reasons which are not fully clear (Newman, 2020; Reuters, 2020). It is speculated that internal migrations were motivated by fears of contagion posed by crowded urban spaces, combined with a desire for more space and larger homes in cheaper locations. In Australia, data shows that average household size increased at the beginning of the pandemic (April to September 2020) but then abruptly reversed following the initial lockdowns, suggesting that pandemic-induced space constrictions may have altered perceptions of personal space, contributing to increased pressure on the already tight property market (Forbes, 2022; Reserve Bank of Australia, 2023).

The stay-at-home orders resulted in worker redundancies, which, in some cases, led to those losing their jobs having to share accommodation with family or friends. They also resulted in people becoming dependent on digital technologies for work, education, and sociality, which seems to have led some people to consider the option of enjoying their 'ideal' lifestyle while continuing to 'work at a distance'. The location and the permanence of these relocations has depended on many factors, of which the nature of the work, possession of requisite financial resources, and access to technologies have been crucial.

In Australia, national statistics released in 2023 suggest that COVID-19 pandemic-induced internal migration may constitute a 'lasting exodus' (Read and McIllroy, 2023). However, by 2022, there was also evidence of reverse-flows of population back to cities as restrictions eased, employers called workers back to the office or factory, and regional areas affected by population increases became less affordable (Bleby, 2023). While the work-from-home and hybrid work trend has intensified, this has varied greatly across different areas of employment and is likely to vary through time according to changes in the economy and range of available employment options (Aksoy et al, 2023). During the pandemic there was much media coverage about internal migrations from cities to suburbs and regions. However, evidence from the United States (US) suggests the enduring appeal of cities even in the midst of the pandemic (Rosanwo, 2021). In the event, there is little doubt that crisis conditions have unsettled established work practices and living arrangements, and reinforced inequalities of work and housing, which has been enabled by and contributed to the advance of technology.

Big Tech and other 'winners' from COVID-19

Big Tech companies and some other types of business benefitted greatly from pandemic emergency conditions, as new digital channels provided new ways to 'connect' with customers, and citizens became dependent on digital technologies and platforms for information about the pandemic and personal risk management measures, conducting work, undertaking education, and performing other activities. A report published by consulting group McKinsey and Company in October 2020 found that 'responses to COVID-19 have speeded the adoption of digital technologies by several years – and that many of these changes could be here for the long haul' (McKinsey and Company, 2020). The report, based on a survey of executives, found that companies had accelerated digitalization of their customer and supply chains and their internal operations by three to four years, and that 'the share of digital or digitally enabled products in their portfolios has accelerated by a shocking seven years' (McKinsey and Company, 2020). Nearly all respondents had made at least temporary solutions to meet the new demands made on them and 'much more quickly than they had thought possible before the crisis'. Moreover, they saw these changes as being sustained and consequently were making the kinds of investments to meet the projected need.

The McKinsey survey revealed that funding for digital initiatives had increased more than increases in costs, the number of workers in technology roles and the number of customers. The report goes on to note that 'Respondents are three times likelier now than before the crisis to say that at least 80 per cent of their customer interactions are digital in nature' and that the rate at which companies were developing these products and services had increased by seven years overall, and by ten years in 'developed Asia'. It was also noted that the increase in digital products was 'much more significant in healthcare and pharma, financial services and professional services' (nearly twice as large) than in the domains of consumer-packaged goods and automotive and assembly products (McKinsey and Company, 2020).

Moreover, the survey results suggest a shift in 'executive mindsets' regarding technology since a similar survey was undertaken in 2017, when 'nearly half of executives ranked cost savings as one of the most important priorities for their digital strategies'. Now, it said 'only 10 percent view technology in the same way; in fact, more than half say they are investing in technology for competitive advantage or refocusing their entire business around digital technologies'. The report concluded by noting that 'the COVID-19 crisis is a tipping point of historic proportions – and that more changes will be required as the economic and human situation evolves' (McKinsey and Company, 2020). Crisis conditions occasioned or accelerated many technological changes, some of which I discuss in the following paragraphs.

Videofication

One significant change during COVID-19 was the rapid integration of videos into people's lives. Video is a technology that calls for a certain kind of affective engagement which advertisers, using AI, have sought to exploit to better 'connect' with potential consumers. While the use of video, including notably platforms such as Zoom and Microsoft Teams, had steadily grown before COVID-19, this greatly accelerated with the onset of national lockdowns in March 2020, when citizens around the world spent extended periods in lockdown and became, in effect, captive audiences for digital media of various kinds. Video consumers became both sources of data and the objects of targeted advertising that exploits this data. As noted in an article published in *Forbes* in November 2021, 'video is reshaping society', and routine interactions previously mostly or exclusively undertaken in person, including meetings, socializing with friends, sales calls, attendance at school, and doctors' appointments, almost instantly moved to video (Bertaux, 2021).

The increase in video use, especially short videos such as employed by TikTok and Instagram, owes much to the development of algorithms that 'mine' massive datasets to detect patterns and, in the case of TikTok, one designed to decide which videos to display to users depending on their personal preferences (Geyser, 2023). TikTok's algorithm responds to a user's behavioural changes to recommend a content that is judged to be appropriate to the moment. Factors that influence the TikTok algorithm include user interactions, information about the video (for example, captions to identify keywords, hashtags to categorize the content, audio such as sounds and songs) and, to a lesser extent, the device and account settings (the device type, country setting, language preference) (Geyser, 2023). An important feature of TikTok's algorithm is that it places weight on whether a user watches a video in its entirety. Consequently, creators who wish to reach a large audience for purposes of advertising are encouraged to create shorter videos that aim to finish within a few seconds. Since the subject matter also holds weight with the TikTok algorithm, producers also need to create content that is targeted to a relevant niche or subculture (Geyser, 2023).

The recent popularity of videos, reflected in the TikTok phenomenon, promises to endure, given its contribution to time-space compression. As Marxist scholar, David Harvey, wrote more than 30 years ago, in capitalist societies, since the early 1970s, there has been 'a sea-change in cultural as well as in political-economic practices … [which is] bound up with the emergence of dominant ways we experience space and time' (1990: vii). While there is no necessary causal connection between time and space, the emergence of new, more flexible forms of capitalist accumulation has led to 'a new round of "time-space compression"' in the organization of capitalism. However, Harvey does not see this time-space compression as a sign of the emergence of an entirely new postcapitalist or post-industrial society, as defying the 'basic rules of capitalist accumulation' (1990: vii).

The advance of digitization and AI, enabled by the harvesting of large datasets in real time, has served in the rearrangement of the relations between time and place. It is no coincidence that investment in the development and application of affective computing has grown rapidly in the wake of COVID-19 and has been predicted to expand at an annual growth rate of 33 per cent between 2020 and 2027 (Grand View Research, 2020). Such data provides the basis for real-time, algorithm-driven advertising using personalized messaging, thereby overcoming the constraints of earlier 'spray and pray' advertising which advertisers now acknowledge is of questionable value in terms of changing behaviours (Petersen, 2023: 37).

Affective computing promises to make advertising more 'personalized' and 'emotionally resonant' and overcome the inefficiencies of the previous 'spray and pray' approach. A major affective computing company, Affectiva, for example, has developed an 'emotion database' derived from 10 million consumer responses to more than 53,000 advertisements in 90 countries over eight years which, it is claimed, will assist advertisers to better engage audiences (Affectiva, 2020). On its website, Affectiva claims that its 'first-of-its-kind emotion database provides unparalleled insights into the impact of emotions in advertising, providing brands with actionable and data-driven advertising direction in our current COVID-19 climate' (Affectiva, 2023) (The company allows users who are willing to share their personal and company details to download 'Mad Men or Sad Men: Trend Analysis Report on Emotions in Advertising', which 'incorporates findings from our extensive research supported by partner data and real ad examples to illustrate our findings' (Affectiva, 2023)).

The COVID-19 pandemic created new opportunities for researchers to use social media to investigate users' emotional experiences, including sentiments shared in social networks. With the sudden surge of users online, following national lockdowns, researchers were able to aggregate text at a global scale to explore mental health and wellbeing, anxieties, collective emotions, and emotional regulation. This psychological research aims to capture shifting sentiments using, for example, multi-lingual dictionary-based methods that analyse certain words that are associated with a particular emotion, assigning weights to words to indicate the strength of the sentiment (Metzler et al, 2022). Machine-learning methods may be used with large databases of words to analyse not just word frequency but word order and other features in context (Metzler et al, 2022).

In Metzler et al's study, the researchers used such methods to analyse Twitter communications before and during the first year of the COVID-19 pandemic which, it is claimed, provided insight into emotional responses during this event, in particular the strong correlation between self-reported sadness and anxiety. Before and during the pandemic, researchers with the Vaccine Confidence Project also tracked public sentiments and emotions relating to current and potential measures to contain and treat COVID-19, undertaking studies in different countries and regions. For example, the Project's website includes details on both quantitative and qualitative studies oriented to understanding the factors, such as social media use, impeding acceptance of vaccines for COVID-19 and other diseases, and approaches to improving confidence in them or countering vaccine hesitancy; for example, using AI and chatbots (Vaccine Confidence Project, 2023).

David Harvey's ideas on time-space compression are helpful for understanding the socio-technological changes in media production and consumption manifest in the increased use of TikTok and other video platforms in recent years. Such advertising is often produced by and/or employs social media influencers who are seen as able to quickly 'connect' with and grow their audiences, often at a global scale, by appearing to be 'authentic' and 'relatable' (Petersen, 2023: 45–50). While Harvey wrote about time-space compression 30 years before COVID-19, when the internet was still in its infancy, the trends he identified have become pronounced with recent advances of digital technologies. Social media and video platforms, such as YouTube (launched in 2005), and more recently TikTok (launched in 2017 and becoming available worldwide after the merger of Chinese social media service, Musical.ly in August 2018), have enabled personal data to be harvested on an unprecedented scale (Petersen, 2023).

In Australia, in 2022 it was reported in the news that a technical analysis by Internet 2.0 had found that 'the Chinese-owned company [that owns TikTok] requests almost complete access to the contents of a phone while the app is in use' (ABC News, 2022). This includes users' personal calendars, contact lists, and photos. It was claimed that the app provided contact to virtually all details on the user's hard drive while the app is in use, as well as all other apps that are stored on the device – more than is needed for the operation of the video-sharing app. This information was being connected to servers around the world, including those in China. A researcher in the company undertaking the analysis noted that the amount of information collected was more than the company said publicly that it claimed to collect (ABC News, 2022). This potential of TikTok to harvest huge amounts of personal data has rapidly increased the platform's value. In January 2022, a Brand Finance

Global 500 report announced that TikTok was 'the world's fastest growing brand', up 215 per cent on the previous year (Brand Finance, 2022).

While in early 2023 the Chinese-owned TikTok was attracting controversy, with many countries banning or threatening to ban it due to concerns about the company's links with the Chinese Communist Party, the short video format that it has popularized seems likely to endure. As explained in *The Economist*, in an article published in March 2023, while 'TikTok's fate hangs in the balance [due to threatened bans in a number of countries] ... what is already clear is that the app has changed social media for good – and in a way that will make it harder for incumbent social apps' (The Economist, 2023). The article continues: 'In less than six years TikTok has weaned the world off old-fashioned social-networking and got it hooked on algorithmically selected short videos. Users love it. The trouble for the platform is that the new model makes less money than the old one, and may always do so' (The Economist, 2023). (The format, it is noted, is less profitable than the news feed.)

The article goes on to note the 'astonishing' speed of change, with the platform in the US attracting more users than other social media apps since entering the country in 2017. Among young users especially, the platform 'crushes the competition', with Americans aged 18–24 spending an hour a day on the platform, which is 'twice as long as they spend on Instagram and Snapchat, and more than five times as long as they spend on Facebook' (The Economist, 2023). The rapid rise of the platform, it is noted, has forced rival companies such as Meta (owner of Facebook and Instagram) to 'reinvent themselves' by turning their apps' main feeds into 'algorithmically sorted "discovery engines"', launching Reels, described as a 'TikTok clone'. Other similar products have been launched by Pinterest (Watch), Snapchat (Spotlight), YouTube (Shorts) and Netflix (Fast laughs) (The Economist, 2023).

In 2021, TikTok had had three billion downloads worldwide, even with a ban in India in June 2020, losing a potentially huge audience. A report from Statista, published in April 2023, noted that while the downloads of the TikTok app had fluctuated between 2021 and the beginning of 2023, the short video platform still ranked as one of the leading apps around the world – considerably ahead of Instagram, WhatsApp, CapCut and Snapchat, respectively (Ceci, 2023a). According to one article, 'TikTok's immense popularity mirrors one of the most significant technological trends of the past decade, the shift from desktop to mobile' and 'reflects recent changes in internet user behaviour, such as shorter attention span resulting in short-form video platforms becoming major players'. It says that this was made evident during the COVID-19 pandemic with 'quality storytelling and accessibility standing out as main reasons to watch short videos' (Ceci, 2023b). TikTok has quickly become a major platform for influencer-based advertising oriented to young consumers, some of which includes products or services or depictions that are harmful; for example, environmentally damaging fast-fashion, beauty products, and body ideals that exploit or reinforce users' insecurities (Bissonette Mink and Szymanski, 2022; Maes and Vandenbosch, 2022; Channel 4, 2023).

YouTube viewing and use of live streaming services also rapidly increased during the first two years of COVID-19. Zoom, which was launched in 2013, proved especially popular with businesses and public organizations during the pandemic. In May 2020, soon after the pandemic began, the easy-to-use platform saw 200 million daily meeting participants; within the next month this rose to 300 million – compared to 10 million in December 2019 (Iqbal, 2023). According to BusinessofApps, 'Zoom was one of the fastest growing apps of the pandemic, and its use for meeting participants increased by 2900%' (Iqbal, 2023). While this platform's level of usage dropped off when many businesses returned to office and schools resumed in-person teaching in 2022, the platform has remained extremely popular in some countries and has become integrated into work and social life although, globally, most usage (about 44 per cent) was by US citizens and predominantly younger (18–54) age groups (Similarweb, 2023).

Yet, Zoom, like many other technology companies, has been affected by post-pandemic inflation pressures, as well as market saturation, with the company announcing in February 2023 that it was cutting 15 per cent of its staff, or about 1,300 people, following a tripling of its staff in two years – effecting what one writer dubbed 'the end of the Zoom boom' (Hoover, 2023). Zoom also faces stiff competition from rivals Microsoft Teams, Google Meet, and Slack, which increased during the pandemic; for example, Microsoft Teams experienced a more than three-fold increase in users between November 2019 and April 2020 (from 20 million to 75 million) (Curry, 2023). Evidently aware that its product needs to offer more than just a video call service, in late 2022 Zoom announced that it planned to integrate e-mail and calendar features into its platform, and employ an AI-driven chatbot to address customer issues, with other features being rolled out in due course (Hoover, 2023).

Visions and applications of AI

As noted, the COVID-19 pandemic created a highly favourable context for the rapid advancement of AI applications. Soon after the pandemic was declared the WHO indicated that AI could be an important technology to manage the crisis (The BMJ, 2023). AI has been critical to pandemic surveillance, including modelling, which I discussed in Chapter 3, and technologies for tracking and tracing infected individuals which have relied on the mining of open source data to develop responses. During the early phases of the pandemic, scientists and policy makers soon focused their attention on the potential of AI to revolutionize pandemic preparedness and response. New pandemic surveillance resources, such as the WHO Hub for Pandemic and Epidemic Intelligence, based in Berlin, sought to harness the power of AI, quantum computing, and other cutting-edge technologies to facilitate greater sharing of data and information between countries and to improve global surveillance for epidemics and pandemics. However, countries were at different stages of integrating AI into their pandemic response (Ghebreyesus and Swaminathan, 2021).

As discussed earlier, COVID-19 also accelerated the advance of AI in the sphere of work. According to a report published by the World Economic Forum in 2021, 'organisations have invested in AI to facilitate remote working, enhance the customer experience and decrease costs' (Reynolds, 2021). In 2020, Appen, a data-collection business, reported that 41 per cent of companies had accelerated their AI strategies as a consequence of COVID-19 (Appen, 2020). Increases in AI use were especially evident in retail (which saw a rise in non-contact pick-up and delivery of meals, reservations to make appointments at retail stores, and the development of more autonomous stores, and the use of computerized models to track inventory); education (where schools and other educational institutions quickly established online programmes and utilized AI-powered smart-learning technologies and virtual tutors and learning software); and healthcare (where the applications have been diverse, including contactless check-in options powered by AI for those needing in-person visits, increased use of telehealth and AI systems for scheduling appointments, and new diagnostic and imaging systems used to diagnose and treat disease) (Reynolds, 2021). I describe some of the many AI-enabled innovations in health and medicine later in the chapter.

While the rapid evolution and application of AI during COVID-19 has been extraordinary, the origins of the technology stretch back to at least the 1950s, when mathematician and computer scientist Alan Turing tested a computer's ability to exhibit intelligent behaviour analogous to that of a human (the so-called Turing test) (Turing, 1950). Turing opened his article with the question, 'Can machines think?' He then wondered whether one could meaningfully interpret and answer this question with conventional methods such as surveys and proposed that it might be replaced by another with 'relatively unambiguous words' using his 'imitation game'. The aim was to test whether the machine (digital computer) could think or reason like people; that is, serve as a 'thinking machine'. The question of whether machines can or could 'think' in the way that humans do continues to be hotly debated, and increasingly so with recent developments in generative AI, especially with OpenAI's release of ChatGPT in November 2022.

Other AI innovations emerging after the onset of COVID-19 include conversational videos produced by US-based StoryFile, which claims to be 'making AI more human' (StoryFile, 2023), 'deadbots' that can simulate conversations with deceased persons (Suárez-Gonzalo, 2022), and 'companion chatbots' such as CarynAI (Lorenz, 2023). Such innovations would appear to reflect a step-change in perceptions of AI and its role in everyday life, in that they suggest that by making it appear more 'human-like' the technology will be able to fulfil people's unmet needs, such as overcoming feelings of loneliness and the constraints of time and space. Yet, while such developments may imply that AI is becoming more 'empathetic' and 'smarter' – perhaps equalling or surpassing human intelligence (the long-imagined state of 'singularity') – it is important to keep in mind that the technology is created by people and that the data that 'powers' its operations reflect human biases, errors, and shortcomings. These biases reflect the outlooks and interests of the people who designed the technologies and of those who contribute, collect, and analyse the data on which they rely.

As highlighted in the 2023 Artificial Intelligence Index Report, produced by Stanford University, research and investment in AI is now led by industry and by men mostly located in the US. Until 2014, most machine-learning models were released by universities, but since 2022 the vast majority have been produced by companies. The report notes that the US leads the world in private investment in AI, with US$47 billion invested by that country, which is about 3.5 times more than is invested in the next highest country, China (US$13.4 billion). Moreover, nearly 80 per cent of new holders of PhD degrees in AI are male (Stanford University Human-Centered Artificial Intelligence, 2023: 11, 18). This report also notes that large language models are 'getting bigger and more expensive'; whereas GPT-2 (an open source AI created by OpenAI), released in 2019, claimed to be the first large language model, had 1.5 million parameters and cost US$50,000 to train, PaLM [Pathways language model, produced by Google], launched in 2022, had 540 billion parameters and cost about US$8 million to train. According to Our World in Data (2023), 'Parameters are variables in an AI system whose values are adjusted during training to establish how input data gets transformed into the desired output'. As the authors of the Artificial Intelligence Index Report note, 'It's not just PaLM. Across the board, large language and multimodal models are becoming larger and pricier' (Stanford University Human-Centered Artificial Intelligence, 2023: 11). The other interesting finding from this report is the environmental impacts of AI, whose training results in substantial carbon emissions; although it is claimed that it can also help 'optimize energy usage' (2023: 73, 120–122).

Since the beginning of the COVID-19 pandemic in 2020, TikTok, Zoom, and other platforms have integrated AI, and specifically generative AI – a type of AI involving machine learning able to generate apparently new content such as images, audio, and text – into their apps. For example, in April 2023 TikTok began experimenting with a new 'profile image generator' which would allow images to produce their own AI profile pictures in-stream. According to one report, 'The generative avatar process needs users to put out between 3 to 10 images of themselves whereas TikTok's system may produce a long range of variations as well as styles for this particular avatar picture' (Anwar, 2023). TikTok, it was noted, 'is on a mission to incorporate its range of AI tools' (Anwar, 2023). Zoom already incorporates AI into their operations, including 'AI-powered features such as virtual backgrounds, avatars, gesture recognition, and background noise suppression' (Parthasarathy, 2023). The platform also uses 'smart recording' using natural language processing, which, as Zoom blog notes, allows users to 'navigate to parts of a meeting that are important to them' (Parthasarathy, 2023).

Google is also investing in a version of generative AI called Bard, as a response to what it evidently sees as a major threat to its long-established position as the internet's main gateway (Liedtke, 2023). The author of the article reporting this development observed that Google's announcement occurred three months after Microsoft's Bing search engine started to utilize the technology similar to that which powers ChatGPT, although the 'AI transition will begin cautiously with the search engine that serves as Google's crown jewel' (Liedtke, 2023). While the technology had initially been available only to people accepted from a wait list, Google claims that it will be available to everyone in more than 180 countries and more languages than English. It is easy to understand why Google would be concerned about competitors like Bing, given that the search engine is the basis for its digital advertising empire which generated more than US$220 billion in revenue in 2022 (Liedtke, 2023).

Notwithstanding growing concerns about data harvesting, data theft, and threats to privacy accompanying increased digitization, hopes for technologies, especially those promising faster, more personalized medical diagnoses and treatments, heightened following COVID-19 and the attention it brought to health and risk.

Digital health innovations

Investment in digital health technologies, which has gained momentum since 2010, skyrocketed during COVID-19 (Meskó, 2022). Many of these technologies have been AI-based. In the US in 2022, medical and healthcare research received the largest single share of AI investment (US$6.1 billion), which was followed by data management, processing, and cloud (US$5.9 billion), and fintech (US$5.6 billion) (Stanford University Human-Centered Artificial Intelligence, 2023: 14). COVID-19 is seen by various commentators to show the technology's potential to offer faster, safer, and more timely and personalized interventions. Emerging and expected applications include the use of AI to assess risks of infection and the screening of populations; remote health monitoring of patients via telehealth; the use of autonomous robots to assist healthcare workers to provide care at a distance; and assisting people to work and learn at home in order to minimize risks of infection, to name but a few.

In hospitals grappling with an increase in COVID-19-related admissions, digital technologies assisted with communications between staff, both within the healthcare system and with those outside the hospital, which was made more difficult by social distancing restrictions. Finding the right staff at the right time is a challenge for any hospital, but especially under pandemic conditions. With the onset of the pandemic, one hospital in Australia was reported to have worked to quickly resolve internal communication issues by introducing an 'ad-hoc, role-based communication solution that allowed doctors in the COVID-19 wards to communicate with doctors outside using Microsoft Teams accounts on dedicated phones which staff would pass on at shift handover'. While this proved difficult to use and could not be scaled the hospital subsequently worked with a health IT company to develop a new role-based messaging solution across the hospital, which was approved by the hospital in March 2022 (Ang, 2023). While such changes may have occurred regardless of the pandemic, the sense of urgency created by emergency conditions led government and business to collaborate in a way that otherwise might have been difficult.

Telehealth gained widespread applications in preventing, diagnosing, treating, and controlling diseases during the COVID-19 outbreak (Monaghesh and Hajizadeh, 2020). The benefits and challenges of telehealth (or telemedicine) have long been debated. However, travel restrictions resulting from lockdowns and border closures in 2020 and 2021 highlighted its many advantages, especially the capacity to incorporate different organizations and situations into a single virtual network. With the onset of COVID-19, virtual care was used for routine, essential medical care, with elective procedures or yearly check-ups being deferred, thereby freeing up medical staff and equipment required to care for those who became seriously ill from COVID-19 (2020: 4). Telehealth also minimized person-to-person contact in waiting rooms and clinics, which was seen to limit the risk of transmission. In Australia, the Federal Government also introduced longer phone consultations for those who had received a positive COVID-19 test, and eligible people were able to have a telephone or video consultation with a doctor to receive a prescription for antivirals (Australian Government, 2022). The healthcare providers allowed to deliver telehealth in Australia during COVID-19 included general practitioners, specialists and consultant physicians, allied health providers, mental health professionals, and nurse practitioners (Australian Government, 2022).

A study examining the extent of telehealth use in Australia, the UK, and the US among those who were frail, living alone, self-isolating, or who needed treatment in the two weeks after the announcement of the pandemic found that while there were differences in practice (for example, the services offered, the extent of attention to older people, the different geographies covered, and funding arrangements), telehealthcare had gathered momentum in the three countries and its acceptance by both patients and health professionals appeared to have increased (Fiske et al, 2020). In Australia, two years after telehealth was introduced by the Federal Government as a temporary pandemic initiative, more than 100 million telehealth services were delivered to 17 million people and more than 92,000 medical practitioners had used the services to support their patients (Department of Health and Aged Care, 2022).

This increased utilization of telehealth in the wake of COVID-19 may partly account for the escalation in the number of reports of mental illnesses during the pandemic. Documented increases in, or worsening of mental illnesses during COVID-19 (Gobbi et al, 2020; World Health Organization, 2022a), including anxiety, depression, and ADHD (Attention Deficit Hyperactivity Disorder), have been attributed by the WHO to 'the unprecedented stress caused by the social isolation resulting from the pandemic' along with 'constraints on individuals' ability to work, and to seek support from loved ones and engage in one's communities' (World Health Organization, 2022b). The pandemic saw a surge of people seeking ADHD assessments at a time when there was a shortage of healthcare providers, creating significant waitlists and leading to the launching of new telehealth services in some countries, making diagnoses of this condition more accessible (Booker, 2022). However, an increase in rates of self-diagnoses leading to formal diagnosis during this period may be attributed in part to people's heightened exposure to social media messages (often posted by celebrities) (Ahmed, 2022).

COVID-19 demonstrated the potential of AI to enable the rapid development of diagnostic tests and vaccines. With any new infectious disease, a priority is to develop a test that can be relied on to quickly identify and isolate affected individuals so that they do not infect others. While the PCR (polymerase chain reaction) tests are considered 'gold standard' in terms of reliably detecting COVID-19, the processing of the results is generally slow and labour intensive, taking hours or days to complete, which can work against a speedy diagnosis required to guide health decisions. The rapid spread of infection in the latter months of 2021 following the emergence of the Omicron variant revealed the limits of PCR-based testing regimes, and many governments changed strategy by relying more on citizens' use of rapid antigen tests, variously known as self-tests, over-the-counter tests, at-home tests, or lateral flow tests (Petersen and Pienaar, 2023). While these tests do not use AI, the idea of self-testing for infectious disease that gained traction during COVID-19 is widely seen by scientists and healthcare professionals to provide a model for future AI-enabled, smartphone-based point-of-care diagnosis (Poirier et al, 2023; Zayed et al, 2023). During the pandemic, AI techniques were employed in studies exploring their potential for diagnosing the disease, which showed high rates of accuracy (Wang et al, 2021). A number of studies found that AI models using deep learning had the ability to discriminate between COVID-19 and non-COVID pneumonia by analysing chest CT (computed tomography) images – although the costs and radiation doses are relatively high (2021: 3). AI tools and machine-learning techniques were also used for analysing the proteome of the virus (its entire complement of proteins) and developing vaccines (Wang et al, 2021).

According to some commentators, AI speeded up research on and the development of vaccines during COVID-19 (Sharma et al, 2022). As Sharma explains, vaccines take on average 15 years to develop, but the pandemic led to a race to develop new ones tailored to COVID-19 (2022: 4–5). AstraZeneca was one of the first companies to use AI in the health sector, and it was used to expedite the detection of biomarkers needed for vaccine development. Pfizer also used AI to run vaccine trials and expedite their distribution (Sharma et al, 2022: 5). Sharma notes that Pfizer began incorporating AI into its working system even before the pandemic, employing it in different aspects of vaccine development and trials (2022: 5).

Growth of data infrastructure during COVID-19

The developments described so far have been made possible by a massive data infrastructure comprising material elements and human labour. This infrastructure includes data brokers who collect and trade data (Melendez and Pasternack, 2019) who, in the US, capitalized on the rise of telehealth for mental health consultations and mental health apps during COVID-19 (Kim, 2023); data centres that store data, and the tens of thousands of behind-the-scenes 'ghost workers' who undertake various forms of 'on-demand', precarious data work (including data moderation, curation, and labelling), generally undertaken from home (Gray and Siddharth, 2019).

In recent years, there has been growing academic interest in platform work, which is both location-based (gig work) and web-based (cloud work), and distributed to individuals or a crowd, respectively (Gruszka and Böhm, 2022: 1858). Both categories of workers are largely invisible to customers and to each other, and even to themselves, but their level of visibility varies in terms of whether they can be seen in the public domain, are institutionally or legally recognized via platform governance mechanisms, and/or subject to algorithmic control and management by platforms; for example, the use of reviews and rating systems and online profiles of workers (2022: 1858–1860).

While this infrastructure may not be visible, it is critical to the operations of digital technologies including AI, and needs to be assembled and maintained, which requires human labour. COVID-19 increased demand for software as a service (the licensing of software on a subscription basis) as growing numbers of people worked from home, which boosted the need for advanced data centres that required specialized power-management technologies (Grand View Research, 2023). Companies own and develop this software using large quantities of data which they purchase from data brokers and/or collect themselves – that is, privatizing and monetizing personal information which citizens have contributed through their online interactions – which is then stored in data centres.

In collecting this data, companies rely extensively on data brokers, of which there are 4,000 worldwide (Kiran and Defensor, 2023). According to one estimate, in 2022 the data broker market was worth US$247 billion and is expected to be worth US$407.5 billion by 2028 (Kiran and Defensor, 2023). Kiran and Defensor (2023) suggest the COVID-19 pandemic has been a factor in this growth. As the authors note: 'For countries under the epidemic [sic] data brokers will help open up public and social data, promote the free flow of data elements, establish new formats of data elements, and give birth to new scenes and occupations.'

Data centres are a critical yet often overlooked component of the digital data architecture. Such centres consume huge amounts of energy and contribute to carbon emissions, and various new technologies have been adopted in the effort to help increase efficiencies. (For an excellent detailed overview of the environmental and social costs of this 'shadow economy' see Pitron, 2023.) The growing use of data centres has also led to the increased use of cloud-based services and mega data centres and colocated data centres by medium-sized enterprises (Grand View Research, 2023). However, while the data centre market has been rapidly increasing – and is expected to grow at 8 per cent annually between 2023 and 2030 – the location of such centres is unevenly distributed around the world. In 2022, North America held the largest revenue share of nearly 38 per cent of data centres globally (Grand View Research, 2023). This concentration of data storage again reflects the inequalities arising from this technological investment, with those living in already resource-rich countries gaining most, leaving those countries least serviced by the internet behind in whatever benefits may derive from technologies.

Change in national science strategies

The onset of the COVID-19 crisis not only spurred research and investment in technologies; it also served to shift national research priorities towards 'health security' utilizing the potential of AI. In May 2023, for instance, the UK Government's Health Security Agency released its Science Strategy 2023 to 2033, which refers to 'a wide range of health security threats' including 'new and emerging infectious diseases, the public health risks posed by climate change, and the growing menace of antimicrobial resistance' (UK Health Security Agency, 2023: 5). The report refers to the Government's concern to 'reduce inequalities and protect the most vulnerable' and states that it will draw on 'data analytics and world-leading science ... to secure major public health victories (such as 'the measles, polio, hepatitis C and HIV elimination targets'), and 'spearhead the UK's efforts to tackle the health security challenges for this decade' (2023: 6).

Not surprisingly, given the hopes for AI already noted, this report refers to the technology's potential to transform medicine and healthcare (2023: 7, 27). Numerous references are made to how COVID-19 responses have shaped research, including in genomics, and how the strategy will build on 'the legacy from the pandemic to strengthen our genomics and surveillance further' (2023: 7). The report notes that 'vaccine discovery, development and evaluation' built on the pre-clinical and clinical development of COVID-19 vaccines 'prevented over 100,000 deaths and enabled the relaxation of other control measures, facilitating socio-economic recovery' (2023: 12, 19). Further, it claims that vaccine development contributed by the UK 'potentially prevented 14.4 million deaths between December 2020 and December 2021'. (2023: 19). (The report announced that the UK Government was building a new Vaccine Development and Evaluation Centre 'that builds on the legacy of the COVID-19 pandemic' (2023: 27)).

Statistics on 'potentially prevented' deaths, such as those cited, are performative in inviting a supportive response, but the ultimate benefits of a national vaccination programme are unknowable. It is interesting that this claim makes no reference to the vaccine nationalism that operated during that period and which denied many low-income countries access to these 'life saving' innovations. There is nothing in the Health Security Agency's report to suggest how exactly science will help address the inequalities that significantly increased during the period of the declared pandemic. While there is a brief reference to 'a greater focus on reducing health inequalities', there are no details about how this will occur (2023: 25). The assumption seems to be that science alone will reduce inequalities, which ignores how science and technology may *increase* inequalities through AI and other innovations that serve to concentrate wealth and advantage only a relatively small proportion of the world's population.

The UK Government's techno-solutionist approach, reflected in the UK Health Security Agency's report, receives support from some influential scientists such as bacteriologist Hugh Pennington, who argues in his 2022 book, *COVID-19: The Post-Genomic Pandemic,* that 'PCR testing on a grand scale has been revolutionary' (2022: 58). Pennington enthuses about how big data ('a fundamental characteristic of the Post-Genomic Age') enabled the rapid, real-time sequencing of the SARS-CoV-2 virus to investigate an outbreak of an infection (2022: 64). Pennington predicts that 'When the next pandemic happens, it will be detected and handled postgenomically' (2022: 108). However, given the evidence emerging from the pandemic regarding increases in social inequalities, the question needs to be asked: who exactly will benefit from this post-genomic 'revolution'?

Conclusion

The pace of socio-technological change in the three years following the onset of COVID-19 has been phenomenal and underlines how a state of emergency, or declared 'crisis', can create conditions propitious for such change. These conditions may have created supposed 'technological wonders', but they have also greatly worsened wealth disparities and generated health and social problems whose full dimensions will likely not be known for many years. The pandemic restrictions have substantially increased reliance on digital technologies and systems, which has advantaged only some people, altered work patterns, and greatly enriched Big Tech and smaller technology companies by enabling data to be harvested and citizens to become experimental subjects on a scale that would have been difficult to imagine before COVID-19.

The advance of AI has been greatly accelerated and looks set to change virtually all aspects of society, including work, law, health, and education, albeit in ways which cannot be predicted. However, it is important to be wary of the hype surrounding new technologies such as AI, since history shows that this is a feature of promising new innovations, which rarely develop in ways envisaged and can be based on a 'false hope' that can both mislead and harm people (Petersen, 2015). In the next and final chapter, I consider what might be learnt from COVID-19 regarding belief in the power of science and technology not just for tackling this pandemic but for future pandemics that may be of equal or even greater magnitude.

6

Future Pandemic Societies

For sociologists and other social scientists, COVID-19 offers valuable insights into how societies respond to a crisis. This event has served, in effect, as a global natural experiment that has shed light on many issues, including the contestations that surround the framing of a public health event as an emergency, the potential for such an event to accelerate trends such as social and technological change, the inequalities that may arise from mitigation measures, and the limitations of risk governance.

In this final chapter I consider what we can learn from this event for societies confronting future pandemics and for public health. I ask whether pandemics could be conceived in ways that are less socially harmful, less stigmatizing, and less polarizing than was witnessed during COVID-19. To help set the scene, it would be useful to begin by examining the issues at stake in the framing of an event as a crisis, in the light of emerging evidence on this pandemic. I then discuss the changes that I believe are needed to better prepare societies to meet the challenges posed by future pandemic disruptions. I conclude with some final remarks on the prospects for social renewal in an age of converging crises.

The framing of a pandemic as a crisis

As I explained in Chapter 1, framing involves attention to certain facts and values and inattention to others, which can limit policy debate on issues and the options for action. The question of how such an event is framed reveals much about societies' values and priorities and the tensions and contradictions inherent in the social order. As discussed in Chapter 2, the UN's and the WHO's initial call for a coordinated response to the COVID-19 pandemic soon revealed the priorities of the G20, which were the interests of its own member countries. It also made evident the danger of relying on experiences of past pandemics and on science-based knowledge to guide responses, given the uncertainties of a novel disease and the responses, as I will explain shortly.

It could be argued that the responses that were made, albeit often delayed and uncoordinated, prevented many illnesses and saved lives; that more people could have been infected and died, had measures such as border closures and lockdowns not been implemented. SARS-CoV-2 proved to be highly contagious, widespread (being reported in 219 countries), and deadly, with more than 2.5 million people losing their lives to the disease in the first year of the pandemic (World Health Organization, 2021d). By end of January 2024, more than 7 million COVID-19 deaths had been reported (World Health Organization, 2024a). In comparison, the bubonic plague, called the Black Death, which ravaged Europe between 1347 and 1351 – widely considered one of the deadliest pandemics in history – was responsible for the deaths of 30 to 50 per cent of Europe's population, according to some estimates (Britannica, 2023). The influenza pandemic of 1918–1919 is believed to have killed between 50 million and 100 million people (Barry, 2004: 397) when the world population was much smaller: 1.8 billion people versus 7.9 billion today.

However, focusing on recorded deaths and/or illnesses, as many health authorities have done when evaluating the impact of COVID-19 and/or the success of interventions, offers a narrow framing of a pandemic and crisis measures. It ignores many issues, including the *longer-term* cascading effects of decisions that cannot be easily measured, such as ongoing economic and social impacts of the disease itself (Long COVID) and mitigation measures. I referred in Chapters 4 and 5 to the inequalities and technological changes set in train by crisis conditions, the full implications of which will likely not be known for many years.

The economic costs of Long COVID alone, measured in terms of reduced earnings, increased medical spending, and lost quality of life, are estimated to be US$3.7 trillion (Cutler, 2022). The personal suffering related to this condition, whose symptoms have been described as similar to myalgic encephalomyelitis or chronic fatigue syndrome (ME/CFS) (Komaroff and Lipkin, 2023), is incalculable. In 2022, the US Centers for Disease Control and Prevention released data from a household pulse survey showed that of the 40 per cent of adults in the US who reported having COVID-19 in the past, nearly one in five (19 per cent) are 'currently still having symptoms of "long COVID"', which are defined as 'symptoms lasting three or more months after first contracting the virus, and that they didn't have prior to their COVID-19 infection' (CDC, 2022).

In March 2023, *The Lancet* reported that 'at least 65 million people are estimated to struggle with long COVID' and that while the majority of those infected with the virus recovered 'within a few weeks', 'long COVID is estimated to occur in 10–20% of cases and affects people of all ages, including children, with most cases occurring in patients with mild acute illness' (The Lancet, 2023). The editors conservatively estimated that 'over 200 million

SARS-CoV-2 infected individuals worldwide will develop persistent symptoms of COVID-19' (Yang and Tebbutt, 2023). Health authorities, including the World Health Organization (2024b), have acknowledged the dimensions and longer-term health implications of Long COVID. However, what are generally not taken into account in discussions about this condition are the ongoing economic, social, and personal impacts, including time off work, resulting in lost income, related medical expenses of treatment, and the strain on personal relationships.

In his statement of 5 May 2023 declaring the end of the pandemic, the WHO's Director-General, Tedros Adhanom Ghebreyesus, acknowledged various impacts of COVID-19, noting that it 'has exposed and exacerbated political fault lines, within and between nations' and 'eroded trust between people, governments and institutions, fuelled by a torrent of mis- and disinformation'. Moreover, 'it has laid bare the searing inequalities of our world, with the poorest and most vulnerable communities the hardest hit, and the last to receive access to vaccines and other tools' (World Health Organization, 2023). This statement provides recognition by the Director-General that the pandemic's impacts go behind 'public health' narrowly defined. However, other comments in the Director-General's statement (which I discuss later), and in other WHO publications, reveal continuing strong confidence in 'science and technology', despite the limitations of science and failures of techno-solutionism during COVID-19.

Growing evidence on the deficiencies of pandemic management, I suggest, should provide the occasion to consider how to reform the global governance of public health to enable societies to better respond to pandemics in the future. What might such reform entail in an age when the international order is undergoing rapid transformation and trust in authorities and science-based expertise has been severely eroded?

Reforming global governance of public health

During COVID-19, the global governance of public health came under considerable strain. The pandemic exposed the difficulties faced by the WHO and the UN in establishing a decisive, coordinated response to a crisis (Chapter 2). Despite their urgent calls for such a response, different countries and jurisdictions reacted with varying degrees of urgency and took approaches that aligned with national or local political priorities and values rather than with what the WHO and UN deemed as essential for 'the public's health'.

In the UK, the government initially prevaricated in its response to the first wave of infection, at first arguing against lockdowns despite the advice of SAGE (Scientific Advisory Group for Emergencies) and then implementing them after evidence from China and Italy made clear the enormous risk posed by the virus (BMA, 2022: 23–24). Sweden, on the other hand, rejected lockdowns on the assumption that the country would achieve 'herd immunity', whereby most people would become infected and develop protection (Aucante, 2022) (see Chapter 1).

Australia, which is considered to have been relatively successful in stemming COVID-19 – purportedly because of its quarantine policy history forged in response to earlier accounts of infectious disease outbreaks (Moloney and Moloney, 2020) – failed to coordinate responses between the different levels of government (Duckett, 2022). The Liberal (conservative) federal government was inclined 'to emphasise the impact of lockdowns on business, and to frame the pandemic response as a trade-off between public health and the economy', and 'showed the tendency of conservative governments … to be suspicious of any restrictions on personal freedoms' (Duckett, 2022: 5).

The crisis offers an opportunity to rethink global preparedness and planning for pandemics; to acknowledge the costs and harms of pursuing different strategies, the inherent uncertainties of pandemics, and the role of emotion-based responses to measures, among other issues that are either ignored in public health or given less attention than they deserve. While the WHO and other supranational organizations expressed concern about preparedness for pandemics as early as 2013 (see Chapter 2), its response to COVID-19 shows that the organization relied too heavily on past experiences of pandemics, which were a poor guide to responding to the uncertainties of the coronavirus and responses to crisis measures.

The danger of historical analogies

As medical historian Robert Peckham argued in *The Lancet* at the beginning of the COVID-19 pandemic, too much focus on the past can constrain one's ability to grasp the complex specific array of factors underpinning the emergence of a disease (Peckham, 2020). Peckham comments that 'When the present is viewed through the lens of former disease outbreaks, we typically focus on similitudes and overlook important differences. In other words, analogies create 'blind spots' (Peckham, 2020). Peckham makes the important point that historical analogies rarely encompass the social and political context within which diseases arise. This context shapes how pandemic crises are defined or classified and consequently what are perceived to be the options for action.

According to the WHO's own definition, a pandemic is 'a worldwide spread of a new disease', and a distinction may be drawn between seasonal and pandemic influenza and the nature and severity of their impacts on different groups (World Health Organization, 2021a). Before 2020, the only WHO-declared pandemics after 1900 were in 1918, 1958, 1968, and 2009 – all of which were influenza (Waltner-Toews, 2020: 10). The WHO's responses to COVID-19 and other recent pandemics appear to have been shaped by expectations based on earlier *influenza* pandemics. For example, the response to pandemic influenza A (H1N1) in 2009 was evidently influenced by experiences of Avian influenza A (H5N1) pandemic, which have different characteristics, the latter being highly lethal but not being easily transmissible between humans; the former, less lethal but highly transmittable between humans (World Health Organization, 2011: 10, 12, 28–29, 38, 84, 104–105, 107, 110, 118).

A report produced by the WHO in 2017, 'Pandemic Influenza Risk Management: A WHO Guide to Inform and Harmonize National and International Pandemic Preparedness and Response', appears to have been an influential guide in its decisions. In the report's Executive Summary, it is stated that 'This guidance introduces a risk-based approach to pandemic influenza management and encourages Member States to develop flexible plans based on national risk assessment, taking account of the global risk assessment conducted by WHO' (World Health Organization, 2017: 9). It goes on to note that, in response to the lessons learnt from the 2009 influenza A (H1N1) pandemic, the guidance was introducing 'a revised approach to global phases ... based on virological, epidemiological and clinical data' (2017: 8). These phases constitute a 'continuum' that the 'Risk Management' document notes 'will be used by WHO to communicate the global situation' (World Health Organization, 2017: 13). They include: 'Interpandemic phase' ('the period between influenza pandemics'); 'alert phase' ('when influenza caused by a new subtype in humans has been identified', calling for 'increased risk assessment at local, national and global levels'); 'pandemic phase' ('the period of global spread of human influenza caused by a new subtype based on global surveillance'); and 'transition phase' (when 'assessed global risk reduces' and 'de-escalation of global actions may occur, and reduction in response activities or movement towards recovery actions by countries may be appropriate, according to their own risk assessments') (World Health Organization, 2017: 13–14). As Waltner-Toews comments, this document appears to be the one that the WHO used during the SARS-CoV-2 pandemic and 'assumes that whenever we are not in a pandemic, we are between pandemics, as we are between ice ages' (Waltner-Toews, 2020: 11).

It is important to note that the classic definition of a pandemic makes no reference to virology or disease severity but, rather, to *the extent of its spread and impact* (Kelly, 2011; emphasis added). A pandemic is defined as 'an epidemic occurring worldwide, or over a very wide area, crossing international boundaries and usually affecting a large number of people' (Last, 2001, cited in Kelly, 2011). As Waltner-Toews notes, 'a disease may become pandemic without being a serious killer', while 'A serious killer disease, even one appearing in different parts of the world, may not necessarily be classified as a pandemic' (Waltner-Toews, 2020: 10).

By this definition, over the last four centuries there has been about one pandemic every 14 years, on average. Of these, there have been an average of 3.8 influenza pandemics per century and an average of 3.4 of other kinds of pandemics; namely, Cholera, HIV/AIDS, Yellow fever, Zika, and COVID-19 (van Bergeijk, 2021: 1). However, over the last century there have been twice as many influenza pandemics as other kinds of pandemic (2021: 1). As Waltner-Toews observes, influenza viruses affect millions of people annually and are called 'seasonal epidemics' but are not labelled pandemics. For the WHO to classify them as a pandemic, there needs to be a global outbreak of a *new influenza A virus*, such as the H1N1 strain linked to the 2009–10 pandemic.

Acknowledging the constraints on WHO

As well as relying too heavily on past experiences of pandemics, the WHO is constrained in its responses by its history, structure, and modus operandi. As noted in the report 'Implementation of the International Health Regulations (2005), Report of the Review Committee on the Functioning of the International Health Regulations (2005) and in relation to Pandemic (H1N1) 2009', the WHO is a servant of its member states [194 in 2021], which meet to set its policy, approve the Organization's budget and plan, and elect its Director-General (World Health Organization, 2011: 11). The Organization's aspirations and responses are conditioned by the various views, needs, and aspirations of its member states. As the report continues, the WHO's response capacities to health emergencies are 'geared towards relatively short, geographically focused events, a type that WHO confronts many times each year' (World Health Organization, 2011: 11). By contrast, the H1N1 pandemic required a worldwide response lasting one to two years (2011: 11). The Review Committee concluded '*The world is ill-prepared for a severe pandemic or for any similarly global, sustained and threatening public health emergency*' (2011: 7; emphasis added).

Given the WHO's conception of pandemics and the various factors that shape its decisions, it is difficult to foresee how it could be reformed in the short-to-medium term. Past recommendations for reform have been largely ignored. The Independent Panel established by the WHO to report on COVID-19, whose findings were discussed in Chapter 4, was scathing in its critique of the Organization. The Panel noted that since the 2009 H1N1 pandemic, '11 high-level panels and commissions have made specific recommendations in 16 reports to improve global pandemic preparedness', with many concluding that the WHO 'needed to strengthen its role as the leading and coordinating organization in the field of health' (The Independent Panel for Pandemic Preparedness and Response, 2021: 16). Further, while some reviews led to specific action, 'the majority of the recommendations were never implemented' and 'pandemics and other health threats have not been elevated to the same level of concern as threats of war, terrorism, nuclear disaster or global economy instability' (2021: 16).

A major factor constraining the operations of the WHO is the conflicting interests of its 194 member states. The WHO, like the UN, emerged as a reaction to a previous crisis, namely the Second World War, and was established to help engender global solidarity in the face of collective threats. However, as Gostin et al (2020) argue, there are many obstacles to global solidarity, with nationalist leaders sometimes seeking to weaken the WHO's authority and weaken a coordinated response. During COVID-19, for instance, member states attacked WHO's leadership and refused to meet their financial obligations to WHO's programming of pandemic responses; notably, the US, which threatened to withdraw entirely from the Organization (Gostin et al, 2020: 1616).

A critical initial step in reforming the global governance of public health is to acknowledge the constraints on the WHO's ability to coordinate responses and develop strategies that assist all nations to better prepare and respond to pandemics (see Chapter 5). As COVID-19 has shown, a pandemic may impact a country with varying intensity, and countries have different capacities to respond. One challenge is 'vaccine nationalism', whereby economically poorer countries may be denied access to vaccines by richer ones during a public health emergency, as occurred during the COVID-19 pandemic (Chapter 4). However, this having been said, it would be wrong for pandemic plans to rely too heavily on science and techno-solutionism and to ignore the socio-political context that may shape responses.

Tempering expectations of science and technology

As COVID-19 has made clear, there is a danger of *expecting too much* of science and technology to tackle pandemics which may involve many uncertainties and unknowns. I am not proposing that one should ignore science-based evidence or adopt the radical sceptical stance of the 'Covid deniers' who believe the pandemic had been fabricated and deaths exaggerated (Shackle, 2021). Rather, I believe that there needs to be greater understanding of the *limits* of science and of the dangers of extensive reliance on technologies to address a highly disruptive, uncertain event. Such an event calls for a multi-pronged approach involving the contributions of different sectors and the many potential stakeholders and publics likely to be affected by measures.

The WHO's declaration of the pandemic in 2020 accorded experts, especially epidemiologists and public health specialists, with extraordinary powers and influence on decisions. Policy makers have relied too heavily on 'science' when justifying decisions even when the scientists disagreed on 'the facts' or expressed concerns about the consequences of those decisions, as was seen with debates about modelling (Chapter 3). The idea that there is or could be agreement among scientists and policy makers regarding 'the evidence' that could inform pandemic strategies in the context of many uncertainties and conflicting political priorities was found wanting.

Scientists disagreed on many issues, including the origin and nature of COVID-19 and the effectiveness and impacts of strategies such as lockdowns and of mass self-testing using rapid antigen tests (Chapter 1). Boundary disputes among scientists, and between scientists and non-scientists, were evident at different stages of the emergency. In science, such disputes are rife, but perhaps especially so with major public health events such as a pandemic, since the stakes are high, with the knowledge and implied practices concerning fundamental issues of life and death and the use of classifications such as who is at risk or poses a risk or is legitimately ill, all of which entail specific expectations, demands, entitlements, and/or exclusions.

While the WHO and most national public health authorities framed the COVID-19 pandemic as an event whose origins, dimensions, and risks could be objectively and comprehensively understood by science and technologically mitigated and managed, few aspects of the pandemic were uncontested. As I noted in Chapter 1, COVID-19 has made clear that there is no agreed reality of a pandemic but, rather, multiple ontologies (to use Annemarie Mol's, 2002 terminology). However, as I have argued, certain realities, namely those of infectious disease epidemiology and public health, dominated conceptions of the pandemic and the framing of responses from its outset. This has had far-reaching consequences for the decisions made and ultimately for people's lives and livelihoods.

With a crisis response, early decisions affect later decisions, creating a path-dependent trajectory that may be difficult to reverse due to political, professional, and business investments in those decisions (see Chapter 1). In the case of COVID-19, these investments included the status and influence of the esteemed experts whose credibility was on the line, the many business owners who put their faith in decision makers to stem infections so that they could 'get back to normal', and ordinary citizens who feared that they or their loved ones would become ill or die from the disease or that they would lose their jobs or businesses (which, for many, has been the case).

The consequences that followed from early decisions – for example, based on modelling employing faulty assumptions, insufficient data, and/or science findings of questionable credibility (concerning, for example, 'herd immunity' (Chapter 1)) – have been far reaching and, in many instances, harmful. In some countries, official framings of the pandemic generated a citizen backlash and a 'crisis of legitimation', leading to a decline of public trust and loss of support for measures. This decline of trust and loss of support, shown in demonstrations of civil unrest (sometimes manifest as violent demonstrations), intensified in some countries (notably China and

Victoria in Australia) as the pandemic dragged on, businesses closed, people lost their jobs, and measures failed to stem infections as new variants emerged.

While the dangers of relying too heavily on science and technology to solve pandemics came to light during COVID-19, the Director-General of WHO, Tedros Adhanom Ghebreyesus, has continued to assert belief in their potential to mitigate pandemics in the future. In his statement declaring the end of COVID-19, made on 5 May 2023, he acknowledged the many deaths, disruptions, and deleterious socio-economic outcomes of COVID-19 but also praised the advances of science and technology. As he explained:

> We have the tools and technologies available to prepare for pandemics better, to detect them earlier, to respond to them faster, and to mitigate their impact. But globally, a lack of coordination, a lack of equity and a lack of solidarity meant that those tools were not used as effectively as they could have been. Lives were lost that should not have been. We must promise ourselves and our children and grandchildren that we will never make those mistakes again. ... If we go back to the way things were before COVID-19, we will have failed to learn our lessons, and we will have failed future generations. This experience must change us all for the better. (World Health Organization, 2023)

These comments suggest that technologies are politically neutral, albeit ineffectively employed artefacts, which raises the question of exactly what 'lessons' the WHO has learnt from COVID-19 and whether there has been a substantial change in the Organization's risk management approach to pandemics. While the Director-General's acknowledgement of the need to avoid repeating mistakes and not to go back to 'the way things were before COVID-19' suggests openness to change, his comments provide no indication of how the noted failures will be addressed. The Director-General's statement makes no reference to WHO's own contribution to these failures and the criticisms made of its handling of the pandemic, discussed earlier.

Acknowledging uncertainty

COVID-19 made clear that a pandemic may involve many uncertainties – about the disease itself, including the manner of infection and paths of transmission, and human conduct – that cannot be predicted by science and the use of mathematical modelling. During COVID-19, science-based risk mitigation was found to be limited in a context of a rapidly evolving pandemic involving many unknowns, including human responses to measures that restricted freedoms of movement and association. Given this, in preparing and planning for pandemics, policy makers need to account for uncertainty, ambiguity, and ignorance surrounding pandemics (Leach et al, 2022). Planning for uncertainty requires acknowledging the various politico-economic and socio-cultural factors that may shape views and actions – factors that cannot be 'modelled'.

During COVID-19, the declaration of the emergency and the demand for policy makers to act quickly to prevent illnesses and deaths and avoid overwhelming hospitals often resulted in a lack transparency in decisions, poor coordination among different actors in the health sector and supply chains, and a failure to garner the views and experiences of the many stakeholder communities potentially affected by decisions (for example, the aged care sector, mental health organizations), among other issues. (For an assessment of national government evaluations of COVID-19 responses, see OECD, 2022.)

COVID-19 showed how 'science' may be used to justify 'risk mitigation/ management' measures that prove costly and harmful (see Chapter 3). Lockdowns, bans on public gatherings, and other restrictions may have reduced the number of infections and deaths and the pressures on hospitals, at least to begin with or at certain stages of a pandemic, but they also exacted huge economic and social costs and disproportionately impacted upon some groups (OECD, 2022). The costs and harms include: the destruction of livelihoods, jobs, and personal finances as many businesses had to close and/or could not trade their goods or services (OECD, 2020b); the deleterious impacts on many children's schooling and, potentially, future careers (for example, Munir, 2021; Rajmil et al, 2021); increased incidences of domestic violence, mental health issues, loneliness, and alcoholism or alcohol-related deaths as people were forced to isolate and/or live alone or in confined spaces for long periods (for example, Piquero et al, 2021; World Health Organization, 2022a, 2022b; National Institutes of Health, 2023); fewer cancer screenings and delayed surgeries as people limited their contact with institutions of healthcare (for example, Luu, 2022; National Cancer Institute, 2022); the direct and indirect social and physical environmental damage resulting from extensive reliance on digital technologies for shopping and other activities (Boudreau, 2021); and increased polarization in public debates and in public demonstrations such as 'lockdown resistance' and related erosion of trust in authorities and their decisions (for example, Sol Hart et al, 2020; Jungkunz, 2021).

The crisis enabled authorities to extend their powers in ways that would have been unacceptable in other circumstances, such as was seen in Melbourne in Victoria, Australia, purportedly the most locked-down city in the world (262 days) between 2020 and 2021 (Chapter 3). (The then Premier, Daniel Andrews, attributes the long period of lockdown to the Federal Government's failure to secure a supply of vaccines as part of its national vaccination plan.) Citizens lost their rights to freedom of movement and to undertake work, and to pursue business and sociable activities, although the suspension of rights played out in different ways in different jurisdictions and countries. Authorities' assumptions that some groups posed a high risk and thus needed to be monitored and/or policed led to them suffering discrimination.

For example, in July 2020, four months after the Australian Government declared the SARS-CoV-2 virus a pandemic, Melbourne saw the State Government impose a lockdown of nine housing commission buildings on the edge of the city where it was believed that the virus was circulating. The lockdown, ordered by the state's top health official, involved hundreds of police surrounding buildings which included many residents from some of the most disadvantaged communities, including those who had escaped war-torn regions such as Somalia and were already distrustful of authorities (Le Grand, 2022: 1–8). One could cite many examples where the extension of authorities' powers enabled or facilitated discrimination against certain groups, causing them considerable harm and in the process undermining trust in the institutions of government.

Given evidence on the many potential deleterious impacts on people's quality of life of harsh lockdowns (as occurred in Melbourne until October 2021, and much of China until December 2022) at least one writer has pondered whether lockdown life was worth living (Lawford-Smith, 2022). It could be argued that the urgency of an emergency allowed little or no time for extensive deliberation on the benefits and harms of lockdowns and other harsh measures. However, given what is known about the dynamics of crises explored in Chapters 1 and 2, many of the aforementioned impacts could have been anticipated and avoided.

The crude utilitarian calculus employed, which placed the 'preservation of health' and potential loss of life above other considerations, including potential harmful outcomes directly or indirectly resulting from measures, has revealed much about societies' values and priorities. The long lockdowns showed disregard for wellbeing as a multidimensional concept – envisaged by the WHO as 'encompass[ing] quality of life and the ability of people and societies to contribute to the world with a sense of meaning and purpose' (World Health Organization, 2021e: 10) – and for established human rights.

Recognizing emotion-based responses

The COVID-19 crisis showed the critical role played by emotions, especially hope, fear, and uncertainty, in shaping policy decisions and personal responses at different stages of its unfolding. Planning for future pandemics needs to pay greater cognisance to how emotions shape actions, including in ways that may not be foreseen by decision makers.

As I discussed in Chapter 1, crises are governed by emotional regimes that call for certain kinds of affective engagement and experience. In preventive health, it is well recognized that the emotions play an important role in shaping public reactions to health communication messages (for example, Gall Myrick, 2015). Health authorities often use fear-laden imagery and messages to encourage audiences to take action to improve their health or to comply with health mandates, such as public health campaigns focusing on HIV/AIDS prevention, smoking, and alcohol abuse – although fear messages are often denounced by activists as stigmatizing (Fairchild et al, 2018). In a pandemic, the generation of fear is likely to encourage adherence to policies, such as handwashing and social distancing (Mertens et al, 2023).

With COVID-19, fear was underpinned by uncertainty about the disease, including its origins, character, risks, and lethality. However, it needs to be recognized that fear may lead to blaming, scapegoating, and stigmatizing groups who are unable or unwilling for whatever reason to conduct themselves in accordance with expert advice, as happened during the pandemic (see Chapter 1). In charting the almost daily rise of cases and sometimes doomsday scenarios provided by modelling (if certain measures were not pursued), epidemiologists and public health experts contributed to creating a climate of fear, and unwittingly contributed to the scapegoating, blaming, and stigmatizing of some groups.

While pandemic measures may offer people hope – the belief and expectation that they will avoid illness and death if they comply with expert advice – during COVID-19 this often proved to be 'false hope' (Rettig et al, 2007). False hope was a consequence of predictions based on incorrect assumptions, and related measures failed to deliver what they promised and/ or had deleterious effects such as those associated with harsh lockdowns, as already noted. Fear and despair were invariably accompanied by periods of hope – a hope that was unevenly distributed between and within populations, with some groups (for example, ethnic minorities, older citizens, hospitality workers, artists of all kinds, younger people in general) and people of the economically poorer nations being marginalized by or effectively excluded from the economy of hope.

Paying cognisance to the inequitable impacts of policies

By mid-2023, more than three years after the declaration of the pandemic, it had become evident that some groups had benefitted greatly financially and in other ways from the crisis (Chapter 4). These groups mostly reside in relatively resource-rich/high-income countries and include technology companies, and those who were able to keep their jobs and build financial wealth through the period of lockdowns or were able to borrow cheap money resulting from governments' stimulus measures to invest in property and other assets. As noted, some companies and individual entrepreneurs were able to capitalize on crisis conditions by successfully exploiting the chaos and gyrations of the stock market (Chapter 1). The pandemic has shown that crisis conditions may create winners and losers. Writers from different fields have done much to expose COVID-19-related inequalities and/or the structural conditions that made these foreseeable; for example, Horton, R (2020a, 2020b), Horton, H (2023), Klein (2020), Caduff (2020), Navarro (2020), Agamben (2021), Desai (2022), and Green and Fazi (2023).

In their edited collection, *Pandemic Exposures*, produced a year into COVID-19, Didier Fassin and Marion Fourcade refer to the uneven and unequal deployment of pandemic relief measures which laid bare the 'stark disparities in life conditions' (2021: 3). As they elaborate: 'People, groups and nations never had equal life chances to begin with, but the pandemic and the turmoil it caused exposed these inequities much more crudely and produced new ones, although along similar hierarchies as the old ones' (2021: 3). The inadequate, inequitable response to COVID-19, the authors argue,

> stemmed from a decades-long erosion of health-care commitments, from the decline in public hospital beds to the delocalization of masks and medicine production, from the evisceration of public health surveillance to insufficient capacities for testing. The crisis was also about the antecedent neoliberal drive to cut both government spending and global production costs. (2021: 3)

As the editors observe, Western media coverage exhibited blindness regarding the plight of places suffering ongoing cataclysms, such as Haiti, Yemen, and Sudan, and the Middle East and Central Africa, which daily grapple with acute respiratory infection, deaths from malaria, famine, conflicts, political chaos, endemic disease, and constant struggle for survival, as well as COVID-19. These problems, they note, did not just disappear with COVID-19 but were 'eclipsed by headline-grabbing tragedies closer to home'. As they write, one of the major characteristics of the pandemic has been the 'narrowing of the economy of attention', which occurred in two ways: 'First, the pandemic, the response to it, the number of cases and deaths became the quasi-exclusive topic in the news and in conversations. Second, interest in the pandemic and its consequences turned inward, for both individuals and nations' (2021: 5).

Most of the issues that Fassin and Fourcade identified in 2021 have been confirmed by the numerous national and supranational enquiries undertaken in 2021 and 2022, including by the WHO-commissioned Independent Panel (Sirleaf and Clarke, 2021) and by the OECD (2022), which has synthesized the evidence of 67 evaluations produced by its member countries during the first 15 months of the pandemic. I examined some of this evidence in Chapter 4.

In their focus on controlling case numbers, public health authorities charged with managing the risks of COVID-19 tended to overlook the wider context within which they worked, including the global and national disparities in wealth and the operations of the media (including social media) that shape how pandemics are represented and responded to. Authorities grappled with managing the representations of the pandemic in the context of the rapidly circulating 'infodemic'. The question is, how might media in its diverse forms (traditional/legacy and social media) be employed to widen debate on issues that are currently mostly neglected in reporting on pandemics, and to counter the misinformation, disinformation, and rumours seen during COVID-19?

Using the media to broaden debate

In Chapter 1, I examined the crucial role played by the news media in framing the COVID-19 pandemic and shaping experiences of and responses to the event. In focusing on only a limited range of issues, the media served to restrict debate on the context within which pandemics arise, assume meaning, and have become increasingly common. Traditional media, such as television and print news, competed with rapidly circulating information of often questionable veracity on the internet. As noted, traditional media provided little coverage of the plight of people in economically poorer countries and the potential measures needed to reduce the prospect of future pandemics, issues that are closely intertwined. The crisis has provided the opportunity to reflect on how the media may contribute to 'widening the lens' on pandemics, to take the predominant focus away from 'downstream' issues of conventional epidemiological interest, such as risk management, infection rates, vaccination coverage, and so on, that dominated during COVID-19, to 'upstream' issues such as the factors that predispose to pandemics, which, many commentators argue are increasing in frequency.

According to the noted US immunologist, Anthony Fauci, societies have entered a new pandemic age in which outbreaks of infectious disease will accelerate as populations grow and climate change intensifies (Vergano, 2020). In reflecting on the impacts of COVID-19, the Australian Nobel Laureate immunologist, Peter Doherty, suggests that societies are witnessing the 'rebirth of the pandemic age', with COVID-19 being but 'a trial run' (Mannix, 2021). Writing during the early phase of the pandemic (in April 2020), Mikkel Frandsen, a Chief Executive Officer of a global social enterprise, makes a similar observation, noting that viral pandemics are 'our new normal'. He says, 'A vaccine next year [2021] will be very welcome, but that is only for this version of Covid-19 let alone the inevitability of new deadly viral pandemics in the years to come' (Frandsen, 2020). If societies are in a pandemic age, as these writers argue, it is crucial that they are positioned to meet the challenges, and it is here that the media can play a critical role in advancing debate on the contributions to recurrent pandemics. A missing part of media reporting on COVID-19, which should be more central to media debate, is the contribution of climate change to pandemics.

According to a recent assessment, climate change and the alteration of ecosystems 'will force new animal encounters – and boost viral outbreaks' (Gilbert, 2022). In an article in *Nature*, Gilbert (2022) suggests that over the next 50 years climate change will alter wildlife habits, increase encounters between species capable of swapping pathogens, and could lead to more than '15,000 new cases of mammals transmitting viruses to other mammals'. There is compelling evidence that climate change is altering the global ecosystem and that societies face a series of 'tipping points', when an accelerating rate of change results in the failure of life-support mechanisms and the emergence of new viruses (Attenborough, 2020: 105–121).

Some researchers have argued that as global temperatures rise there is an increased likelihood that viruses and bacteria that have been locked up in glaciers and permafrost may 'reawaken' and infect local wildlife (for example, Geddes, 2022; Strona et al, 2023; Varghese et al, 2023). According to one estimate, 58 per cent (218 of the 375) of infectious diseases encountered by humans have been exacerbated by climatic changes (Varghese et al, 2023). These researchers claim that there is empirical evidence showing that there are '1006 different ways in which climate change paves the way for a spectrum of diseases' (Varghese et al, 2023). A 2016 outbreak of anthrax in northern Siberia that killed a child and infected at least seven other people has been 'attributed to a heatwave that melted permafrost and exposed an infected reindeer carcass' (Geddes, 2022). A news report on this case indicated that the last anthrax outbreak in the region was in 1941 (Geddes, 2022).

Researchers who have been collecting and analysing soil and sediment samples from Lake Hazen in Canada, which is fed by meltwater from local glaciers, used RNA and DNA sequencing to 'identify signatures closely matching those of known viruses, as well as potential animal, plant, or fungal hosts, and ran an algorithm that assessed the chance of these viruses infecting unrelated groups of organisms' (Geddes, 2022). In the article reporting this finding, the researchers noted that while the risks of this 'viral spillover' are difficult to quantify, climate change is rapidly transforming environments in ways that increase the likelihood of such an outcome (Lemieux et al, 2022). They also suggest that 'Should climate change also shift species range of potential viral vectors and reservoirs northwards, the High Arctic could become a fertile ground for emerging pandemics' (Lemieux et al, 2022).

Important sociological insights may be gained from analysing how pandemics are framed in media and public policies and the consequences of this for both decisions and 'non-decisions', or 'the practice of limiting the scope of actual decision-making to "safe" issues by manipulating the dominant community values, myths, and political institutions and procedures' (Bachrach and Baratz, 1963: 632). In respect to climate change, the well-known sociologist Bruno Latour (who died in October 2022) saw the experiences of the COVID-19 lockdowns as both interesting and empowering, reminding humans that they are 'terrestrials living on earth' having shaken off 'the secularised religions that preach escape from this world' (Latour, 2021: 58). Latour's book serves as a wake-up call, alerting humans to the fact (evidently largely forgotten) that they are part of a wider ecosystem and have a responsibility to the planet which they have helped to warm through carbon emissions contributed to by capitalist processes of production and consumption. The latter includes digital technologies and increasingly sophisticated artificial intelligence (AI) tools, which consume huge amounts of electricity and natural resources, thereby contributing to carbon emissions (Pitron, 2023).

Implementing measures to reduce or mitigate the impacts of climate change, however, is extremely difficult, given the relentless pursuit of 'growth' and strong belief in what technologies, especially AI, will deliver in the future. The neoliberalism that dominates political rule in many societies has meant governments are ascribed a limited role in taking measures that will reduce carbon emissions. From the outset of COVID-19, authorities have sought to strike a balance between the tensions and contradictions arising from the simultaneous pursuit of economic goals, on the one hand, and the preservation of 'the public's health' and security, on the other. This tension has been reflected in the crude binary debates about 'saving jobs versus saving lives' when the lockdowns were first implemented in 2020, which suggests that these are opposed, incompatible objectives (Chapter 1). COVID-19 in fact made transparent the *inextricable links between health and economy*, in that the health impacts were exacerbated and compounded by an unsustainable economic globalization involving 'the neoliberal dismantling of state capabilities in favour of markets' (van Barneveld et al, 2020). As has often been the case during the pandemic, technologies have been portrayed as the solution to these conundrums rather than as a contributor to them.

As I argued in previous chapters, particularly Chapter 2, the COVID-19 pandemic has demonstrated the limitations and failures of 'risk governance'. At each stage of the pandemic, socio-political factors have shaped decisions regarding interventions for stemming the spread of infections and minimizing disease and deaths. In this 'maths of life and death' (Yates, 2021) scientists have made judgements based on a risk management approach that has had far-reaching impacts on people's lives. This includes the development and adoption of new technologies during COVID-19, such as AI and genomic sequencing, and changes in how science is practised (Chapters 3 and 4). Yet, despite the rapidity of recent developments, there is nothing inevitable about the direction of technological change. There are always alternative options for action – ones that are less expert driven and technology reliant – which can contribute to the creation of societies that are more equal, healthier, less polarized, and more compassionate and caring than those revealed during the COVID-19 pandemic.

In the following final paragraphs, I consider the prospects for such change provided by the conditions of crisis.

Polycrisis and prospects for social renewal

A common saying, variously attributed to Winston Churchill, Italian Renaissance philosopher Niccolo Machiavelli, and Rahm Emanuel, the chief of staff to President-elect Barack Obama, is 'Never let a good [or serious] crisis go to waste'. The saying reflects the belief that a crisis creates the opportunity to do things that one could not otherwise do. This opportunity-creating potential of crises has long been recognized by sociologists and other scholars, including one of the discipline's recognized founders, Emile Durkheim, whose work I discussed in Chapter 2. A crisis illuminates matters that are invisible most of the time, since they are a taken-for-granted aspects of the social order. As noted, the disruptions of crises may lead people to see the world differently and provide the impetus to address matters that have been neglected.

A sense of urgency for change has been reinforced by the concurrence of various catastrophic events – a so-called polycrisis; namely, the 'interplay between the COVID-19 pandemic, the war in Ukraine and the energy, cost-of-living and climate crises' (Whiting and Park, 2023). In 2024, one can add to this list: the war in the Middle East and the related political polarization, the crisis associated with recent rapid technological advances, especially in AI and algorithm-driven processes, and the crisis of trust in authorities and expertise in many societies. These crises do not just occur simultaneously but are mutually reinforcing. Their concurrence provides an opportunity for rethinking and reframing pandemics (including public health's conception of them) to account for the socio-political factors that shape their representation and the responses to them. As Baker et al (2022) argue, a rapidly changing world requires a changing science to evaluate risks from infectious disease. This calls for research that enables societies to contend with possible future outcomes, in addition to the retroactive analyses that dominate the literature, with attention paid to (among other crisis-related changes) the emergence of pathogens associated with climate change and the encroachment of agriculture into native species' habitats and invasive species moving into populous regions (Baker et al, 2022: 203).

As I noted in Chapter 1, much has already been written on COVID-19 and its consequences, including by sociologists and other social scientists who have examined various responses to and short-term impacts of this event. The avalanche of research undertaken during the crisis (Pai, 2020) has produced many valuable insights. However, as argued in Chapter 1, it has tended to focus on relatively short intervals of time; in effect, creating snapshots of a rapidly evolving, uncertain phenomenon. This has meant that matters that seemed critical or to have particular significance at one point in time have been subsequently re-evaluated and found to be less important or to assume a new meaning.

For example, in the UK in 2020, experts' fears about the potential for hundreds of thousands of COVID-19-related deaths and the collapse of the National Health Service if a national lockdown was not implemented – based on modelling and observations of high mortality rates and impacts on intensive care facilities in northern Italy at the time – were later judged to be ill founded, since the models employed faulty assumptions (Chapter 3). Interventions designed to 'save economies' and to minimize pandemic-related disruptions, implemented in 2020, have been subsequently found to have fuelled inflation which has reduced the living standards of many people (Chapter 4). The impacts of many pandemic measures, including supply chain disruptions, labour shortages, inflationary pressures, increased wealth disparities, and educational and emotional harms, came to light in only the second or third year following the declaration of the emergency.

Public health needs to be reconceptualized to encompass issues and perspectives that have received insufficient attention or been ignored during the COVID-19 pandemic, such as those discussed earlier. However, developing public health approaches that will better equip societies to prevent or minimize the prospect of pandemics and/or to mitigate and manage their effects is difficult, given the dominance of epidemiological and risk management approaches.

As Deborah Lupton and I argued nearly 30 years ago, epidemiology is integral to the history and identity of public health with its focus on the measurement and classification of populations and statistical analyses of 'scientific facts', especially the 'factors of risk' (Petersen and Lupton, 1996: 28–35). While the origins of epidemiology, the study of epidemics, can be traced back to the 19th century, it became an academic field of study in the decades following the Second World War (Oppenheimer, 1995: 918). The emergence of this specialism followed an 'epidemic transition', whereby chronic, non-infectious disorders (such as heart disease and cancer) displaced infectious diseases as a major cause of mortality in resource-rich Western countries (1995: 918).

In the 1990s, epidemiology, like public health more broadly, largely focused on the promotion of 'healthy lifestyles', the sources of personal risk, and risk management practices – what was known as 'the new public health'. Over the last two decades or more, public health has paid much less attention to the reality of infectious diseases, which many public health experts assumed had become less prevalent than chronic, lifestyle-related diseases of modern living. This assumption reflects the widely held view that 'science' has largely 'won the war' against infectious diseases; an assumption that has contributed to the neglect of the plight of those living in poorer countries, who daily grapple with both communicable and non-communicable diseases, and the failure to envisage the potential emergence of pandemics.

To be sure, during the pandemic, some individual scientists and practitioners *did* see the wider context within which they worked, and questioned the assumptions of science, and/or reliance on certain technology-based interventions, but they were in the minority (Chapter 3). For example, some scientists questioned the strategy of mass self-testing employing rapid antigen tests (RATs) (also called lateral flow tests or at-home tests), with specialists raising concerns about the use of RATs when they were first granted 'emergency use' approval in the US in August 2020. These specialists were concerned that the self-tests would miss infectious people, resulting in outbreaks in countries that had largely controlled COVID-19 transmission, and that errors would arise from people using different criteria to make judgements on their potential infectiousness. Some argued that the strategy was misguided and unlikely to reduce transmission (for example, Guglielmi, 2020; Raffle and Gill, 2021). (For a more detailed discussion on the complexities of implementing Australia's mass self-testing strategy during COVID-19 see Petersen and Pienaar, 2023.)

Yet, while some individual scientists and some groups of scientists raised concerns about certain models or the effectiveness or the health and/ or social impacts of particular strategies – for example, those involved in the Great Barrington Declaration (see Chapter 3) – few questioned the dominant framing of the pandemic or the premises of epidemiology and risk management, or the utilitarianism underlying decisions, even when interventions failed to fulfil their promise or produced demonstrable harms or met strong community resistance or contributed to social unrest.

One high-profile scientist who did question how the COVID-19 pandemic was framed is Richard Horton, editor-in-chief of *The Lancet* (mentioned in Chapter 2), who appears to have quickly changed his position on COVID-19 as the pandemic evolved. On 24 January 2020, Horton tweeted: 'A call for action please. Media are escalating anxiety by talking of a "killer virus" + "growing fears". In truth, from what we currently know, 2019-nCOV has moderate transmissibility and relatively low pathogenicity. There is no reason to foster panic with exaggerated language' (Science Media Centre, 2020). In September 2020, however, Horton published an article in *The Lancet* where he argued: 'As the world approaches 1 million deaths from COVID-19, we must confront the fact that we are taking a far too narrow approach to managing this outbreak of a new coronavirus' (Horton, 2020a). In his view, 'The "science" that has guided governments has been driven mostly by epidemic modellers and infectious disease specialists, who understandably frame the present health emergency in centuries old terms of plague' (Horton, 2020a).

Horton offered a novel perspective on the pandemic, arguing that COVID-19 should be understood as 'Two categories of disease interacting within specific populations – infection with severe acute respiratory syndrome coronavirus 2 (SARS-CoV-2) and an array of non-communicable diseases (NCDs)'. He added: 'These conditions are clustering within social groups according to patterns of inequality deeply embedded in our societies' (Horton, 2020b). Consequently, he argued, 'COVID-19 is not a pandemic [but rather] a syndemic', which required a 'more nuanced approach ... to protect the health of our communities' (Horton, 2020a). Horton outlined his concerns about the global response to the pandemic in his book, *The COVID-19 Catastrophe: What's Gone Wrong and How to Stop It Happening Again* (Horton, 2020b).

Horton's apparent change of position from scepticism about the danger posed by the media-generated panic about the virus to criticism of the 'too narrow approach to managing this outbreak of a new coronavirus' and his assessment of COVID-19 as being 'syndemic' which requires a 'more nuanced approach', shows that experts *may* change their views when confronted by new evidence or arguments. Other experts also raised concerns about reliance on computational models, especially the pressure to respond to policy makers' demands for rapid advice despite the absence of adequate, reliable data and the failure of responses to account for the behavioural and social complexity of societies in a context of pandemic crisis (for example, Squazzoni et al, 2020). (See also Chapter 3.)

Yet, expert questioning during COVID-19 has generally had more to do with 'boundary disputes'; namely, who can legitimately lay claim to define the situation. For example, in the UK, in September 2020, two groups of prominent scientists wrote open letters to the government with conflicting advice on restrictions. One group of 32 scientists signed a letter that warned that the government should reconsider its policy to suppress the virus and instead adopt a 'targeted approach'. At the same time, another group of 23 scientists wrote to the Chief Medical Officer and the Chief Scientific Advisor, apparently supporting the government's approach which involved suppression across the entire population (Buranyi, 2020b). A news report at the time stated that 'The two stances underline a schism within the scientific community over how to tackle the second wave of coronavirus in the UK' (Buranyi, 2020b). While such schisms are not uncommon in science communities, they are inclined to play out in public forums in situations of crisis when the stakes are high and demands for urgent policy decisions bring conflicts to a head.

As the COVID-19 pandemic has shown, pandemics are inherently unstable realities, and perhaps more so in the age of digital communications. The advent of social media and generative AI, which may be used by state and non-state actors to create and disseminate diverse, sometimes unreliable information – such as that which is falsely attributed to political leaders or credentialed experts – means that authorities faced with pandemics in the future will find it even more difficult than in the past to control their representations. In May 2023, a group of AI experts released a 'Statement on AI Risk', which proposed that 'AI should be a global priority alongside other societal-scale risks such as pandemics and nuclear war' (Center for AI Safety, 2023).

The advance of AI increases the potential for proliferating narratives on pandemics to destabilize official framings and communications. COVID-19 revealed many competing conceptions of issues such as safety (for example, as occurred with AstraZeneca vaccines and claims about their potential to create blood clots), herd immunity, the dangers posed by new virus variants, the effectiveness of vaccines in general or certain kinds of vaccine, and the question of who poses a greater risk of spreading infection or is more at risk of disease, among other issues. This competition to frame issues will very likely intensify as AI tools become embedded in everyday life and more and more people exploit their affordances.

Conclusion

Critical scholars, including sociologists and other social scientists can play a crucial role in exposing and analysing the multiple ontologies of pandemics and the implications for future societies confronting recurrent infectious disease outbreaks. This includes the polarization of views that may be contributed to by pandemic responses that had been evident from the beginning of the COVID-19 pandemic. However, in undertaking this critical work it is important that scholars reflect on the blind spots in their own disciplines, especially concerning the plight of those living in economically poorer parts of the world.

As Connell argues, while social science has a vital role to play in democracy, it is 'at best, ambiguously democratic' (2007: vii). She elaborates:

> Its [social science's] dominant genres picture the world as it is seen by men, by capitalists, by the educated and affluent. Most important, they picture the world as seen by the rich capital-exporting countries of Europe and America – the global metropole. To ground knowledge of society in other experiences remains a fragile project. Yet only knowledge produced on a planetary scale is adequate to supporting the self-understanding of societies now being massively reshaped on a planetary scale. (2020: vii)

Connell is concerned with 'how social science might operate democratically on such a scale'.

Critical scholars can contribute to this democratic project by shifting the focus away from viewing pandemics as mostly or purely biophysical phenomena that call for epidemiological and public health expertise, technologies, and technologies, to consider the politico-economic and socio-cultural factors that predispose to their occurrence and shape the responses. Through their research and writing scholars can help uncover taken-for-granted assumptions that inform and serve to narrow public discourse – especially those of epidemiology and risk management – and turn attention to a broader range of issues that are currently mostly neglected in pandemic preparedness and planning.

Topics calling for such analysis include: the potential of emergency measures to reinforce inequalities and polarization; over-reliance on outdated approaches and on technologies for making decisions; media's role in narrowing attention and thinning debate, especially regarding the plight of those in economically poorer countries whose living conditions, often in crowded megacity slums, contribute to the rise and spread of infectious disease (Muggah and Florida, 2020); and the potential harms resulting from the exercise of a simplistic utilitarian logic involving trade-offs between the hypothetical 'benefits' and 'harms' of pursuing different measures (Chapter 3).

The project of rethinking pandemics should, above all, include consideration of the forms of governance most likely to advance the health and security of all citizens. This governance would include mechanisms for strengthening of the bonds of civil society that have been severely eroded over more than four decades of neoliberal policies that rely on market-based mechanisms to resolve complex socio-political problems and shift responsibility onto citizens for matters that were previously the responsibility of the state or not responsibilities at all (O'Malley, 2009).

In short, the COVID-19 pandemic has provided the opportunity to reimagine societies – ones that are very different from those promised by neoliberalism, 'risk management', and techno-solutionism. By contributing to this project, critical scholars can help advance the project of social renewal, which is crucial not only to enable societies to be less vulnerable to future pandemics and their disruptions but also to avoid the harmful outcomes of responses that were all too evident during COVID-19.

References

ABC News (2022) 'Video: TikTok users warned the platform is harvesting personal data', 18 July. www.abc.net.au/news/2022-07-18/tiktok-users-warned-the-platform-is-harvesting-personal-data/13977370 (Accessed: 27 April 2023)

Achenbach, J. (2023) 'What we know about the origin of covid-19, and what remains a mystery', *Washington Post*, 28 February. www.washingtonpost.com/science/2023/02/28/covid-origin-evidence/?utm_campaign=wp_post_most&utm_medium=email&utm_source=newsletter&wpisrc=nl_most&carta-url=https%3A%2F%2Fs2.washingtonpost.com%2Fcar-ln-tr%2F393f56b%2F63fe393f1b79c61f87ae9c9b%2F5e86729bade4e21f59b210ef%2F41%2F74%2F63fe393f1b79c61f87ae9c9b&wp_cu=30407fe69d6a6394f230585c479f343f%7CA25801DC315965F6E0530100007F2D9A (Accessed: 2 March 2023)

Affectiva (2020) 'Research data shows advertising is having a more emotional impact on consumers, but not all advertisers are using emotions effectively'. www.businesswire.com/news/home/20201027005291/en/Research-Shows-Advertising-is-Having-a-More-Emotional-Impact-on-Consumers-But-Not-All-Advertisers-Are-Using-Emotions-Effectively (Accessed: 4 May 2023)

Affectiva (2023) 'Mad men or sad men: trends analysis report on emotions in advertising'. https://go.affectiva.com/trends-in-advertising-insights-report (Accessed: 4 May 2023)

Agamben, G. (2005) *State of Exception*. Translated by Kevin Attell. The University of Chicago Press: Chicago and London.

Agamben, G. (2021) *Where Are We Now? The Epidemic as Politics*. 2nd updated edn. Translated by Valeria Dani. Eris: London.

Agassi, J. (2003) *Science and Culture*. Kluwer Academic Publishers: Dordrecht.

Ahmed, T. (2022) 'Psychiatrist on what is to blame for rising ADHD rates', *News.com.au*, 16 December. www.news.com.au/lifestyle/health/health-problems/psychiatrist-on-what-is-to-blame-for-rising-adhd-rates/news-story/d6e064992a68e325b62706b89004acf7 (Accessed: 30 May 2023)

Aksoy, C.G., Barrero, J.M., Bloom, N., Davis, S.J., Dolls, M. and Zarate, P. (2023) 'Working from home around the world', Centre for Economic Performance Discussion Paper No 1920, May. https://cep.lse.ac.uk/pubs/download/dp1920.pdf (Accessed: 14 July 2023)

Altman, S.A. and Bastian, C.R. (2021) 'The state of globalization in 2021', *Harvard Business Review*, 18 March. https://hbr.org/2021/03/the-state-of-globalization-in-2021 (Accessed: 2 May 2023)

Anderson, A., Petersen, A., Wilkinson, C. and Allan, S. (2009) *Nanotechnology, Risk and Communication*. Palgrave Macmillan: Houndmills and New York.

Anderson, W. (2021) 'The model crisis, or how to have critical promiscuity in the time of Covid-19', *Social Studies of Science*, 51, 2: 167–188.

Andrews, T.M. (2020) 'Facebook and other companies are removing viral "Plandemic" conspiracy video', *Washington Post*, 8 May. www.washingtonpost.com/technology/2020/05/07/plandemic-youtube-facebook-vimeo-remove/ (Accessed: 12 June 2020)

Ang, A. (2023) 'Covid-era solution enables seamless communication among almost 6,000 hospital staff', *HealthcareITNews*, 15 May. www.healthcareitnews.com/news/anz/covid-era-solution-enables-seamless-communication-among-almost-6000-hospital-staff?mkt_tok=NDIwLVlOQS0yOTIAAAGLxixQVOAcMbqBnvgPVl3VdGvU1ekcNvmNlWzirKThGiPJM9AIy-NHgM6GRBM5dPoNWT_mRXmuB7UpEuCUGj-TqCQcyHN4624PhiJk6mCUGA (Accessed: 23 May 2023)

Anirudh, A. (2020) 'Mathematical modeling and the transmission dynamics in predicting the Covid-19: what next in combating the pandemic?', *Infectious Disease Modelling*, 5: 366–374.

Anwar, H. (2023) 'TikTok begins testing its new generative AI avatar creator', *Digital Information World*, 26 April. www.digitalinformationworld.com/2023/04/tiktok-begins-testing-its-new.html (Accessed: 2 May 2023)

Appen (2020) 'The 2020 state of AI and machine learning report', 21 May. https://appen.com/whitepapers/the-state-of-ai-and-machine-learning-report/ (Accessed: 18 May 2023)

Arendt, H. (1951) *The Origins of Totalitarianism*. Harcourt, Brace and Co: New York.

Arendt, H. (1977) *Between Past and Future: Eight Exercises in Political Thought*. Penguin Books: London.

Asai, T. (2023) 'Pandemic and infodemic: the role of academic journals and preprints', *Journal of Anesthesia*, 37, 2: 173–176.

Attenborough, D. (2020) *A Life on Our Planet: My Witness Statement and a Vision for the Future*. Witness Books: London.

Aucante, Y. (2022) *The Swedish Experiment: The COVID-19 Response and Its Controversies*. Bristol University Press: Bristol.

Australian Government (2022) 'Proving health care remotely during the COVID-19 pandemic: telehealth services'. www.health.gov.au/health-alerts/covid-19/coronavirus-covid-19-advice-for-the-health-and-disability-sector/providing-health-care-remotely-during-the-covid-19-pandemic (Accessed: 29 May 2023)

Australian Government (2023) *Statistical Report. Aviation: Domestic Aviation Activity January 2023*. Department of Infrastructure, Transport, Regional Development, Communication and the Arts: Canberra. www.bitre.gov.au/sites/default/files/documents/domestic-aviation-activity-publication-january-2023.pdf (Accessed: 19 April 2023)

Azcona, G., Bhatt, A. and Kapto, S. (2021) 'The COVID-19 boomerang effect: new forecasts predict sharp increases in female poverty', UN Women, Research highlight, gender and Covid-19. https://data.unwomen.org/features/covid-19-boomerang-effect-new-forecasts-predict-sharp-increases-female-poverty (Accessed: 28 April 2021)

Bachrach, P. and Baratz, M.S. (1963) 'Decisions and non-decisions: an analytic framework', *The American Political Science Review*, 57, 3: 632–642.

Baker, R.E., Mahmud, A.S., Miller, I.F., Rajeev, M., Rasambainarivo, F., Rice, B.L., Takahashi, S., Tatem, A.J., Wagner, C.E., Wang, L-F., Wesolowski, A. and Metcalf, J.E. (2022) 'Infectious disease in an era of global change', *Nature Reviews/Microbiology*, 20 (April): 193–205.

Barry, J.M. (2004) *The Great Influenza: The Story of the Deadliest Pandemic in History*. Penguin: London.

BBC News (2020) 'Coronavirus: fake news crackdown by UK government', 30 March. www.bbc.com/news/technology-52086284 (Accessed: 17 July 2020)

BBC News (2021a) 'Wuhan lab leak theory: how Fort Detrick became a centre for Chinese conspiracies', 23 August. www.bbc.com/news/world-us-canada-58273322 (Accessed: 7 September 2021)

BBC News (2021b) 'Covid: regulator criticises data used to justify lockdown', 5 November. www.bbc.com/news/health-54831334 (Accessed: 2 February 2024)

Becker, G. (1997) *Disrupted Lives: How People Create Meaning in a Chaotic World*. University of California Press: Berkeley, CA.

Bell, K. and Green, J. (2020) 'Premature evaluation? Some cautionary thoughts on global pandemics and scholarly publishing', *Critical Public Health*, 30, 4: 379–383.

Bellware, K., O'Grady, S., Shaban, H. et al (2020) 'Live updates: US sets record for new coronavirus cases, surpassing 53,000', *Washington Post*, 3 July. www.washingtonpost.com/nation/2020/07/02/coronavirus-live-updates-us/ (Accessed: 3 July 2020)

Belot, H. (2023) 'Flights of fancy: why Australians are spending big on private jets', *Guardian*, 16 April. www.theguardian.com/australia-news/2023/apr/16/flights-of-fancy-why-wealthy-australians-are-spending-big-on-private-jets (Accessed: 19 April 2023)

Bertaux, G. (2021) 'How video is reshaping society in the wake of the pandemic – and what to expect for 2022', *Forbes*, 29 November. https: www.forbes.com/sites/forbestechcouncil/2021/29/how-video-is-shaping-society-in-the-wake-of-the-pandemic--and-what-to-expect-for2 022/?sh=2e109fac7508 (Accessed: 27 April 2023)

Bessen, J. (2022) *The New Goliaths: How Corporations Use Software to Dominate Industries, Kill Innovation, and Undermine Regulation.* Yale University Press: New Haven, CT.

Bissonette Mink, D. and Szymanski, D.M. (2022) 'TikTok use and body dissatisfaction: examining direct, indirect, and moderated relations', *Body Image*, 43: 205–216.

Bleby, M. (2023) 'Why CBD building will become "recruiting machines"', *Financial Review*, 9 January. www.afr.com/property/commercial/why-cbd-buildings-will-become-recruiting-machines-20221215-p5c6hj (Accessed: 11 May 2023)

Bloomberg (2021) 'More than 1.06 billion shots given: Covid-19 tracker'. www.bloomberg.com/graphics/covid-vaccine-tracker-global-distribution/#global (Accessed: 28 April 2021)

Bogle, A., Workman, M. and Hutcheon, S. (2020) 'How coronavirus "changes the game" for the anti-vaccination movement', *ABC News*, 31 May. www.abc.net.au/news/2020-05-31/anti-vaxxers-are-exploiting-the-coronavirus-crisis/12302710?nw=0 (Accessed: 12 June 2020)

Bonotti, M. and Zech, S. (2021) *Recovering Civility during COVID-19.* Palgrave Macmillan: London.

Booker, Z. (2022) 'Online treatment for ADHD: what telehealth companies should know', Forbes, 18 April. www.forbes.com/sites/forbesbusinesscouncil/2022/08/18/online-treatment-for-adhd-what-telehealth-companies-should-know/?sh=2630748b9ba1 (Accessed: 30 May 2023)

Boudreau, C. (2021) 'Shopping online surged during Covid. Now the environmental costs are becoming clearer', *Politico*, 18 November. www.politico.com/news/2021/11/18/covid-retail-e-commerce-environment-522786 (Accessed: 2 February 2024)

Bowker, G.C. and Star, S.L. (2000) *Sorting Things Out: Classification and Its Consequences.* Cambridge, MA: The MIT Press.

Box, G.E.P. (1976) 'Science and statistics', *Journal of the American Statistical Association*, 71, 356: 791–799.

Brainard, J. (2021) 'No revolution: COVID-19 boosted open access, but preprints are only a fraction of pandemic papers', *Science*, 373, 6560: 1182–1183.

Brainard, J., Rushton, S., Winters, T. and Hunter, P.R. (2020) 'Introduction to and spread of COVID-19-like illness in care homes in Norfolk, UK', *Journal of Public Health*. DOI: 10.1093/pubmed/fdaa218.

Brand Finance (2022) 'TikTok named world's fastest-growing brand', 26 January. https://brandfinance.com/insights/brands-are-bouncing-back (Accessed: 28 April 2023)

Brauer, F. (2017) 'Mathematical epidemiology: past, present, and future', *Infectious Disease Modelling*, 2: 113–127.

Bringel, B. and Pleyers, G. (2022) *Social Movements and Politics during COVID-19: Crisis, Solidarity and Change in a Global Pandemic*. Bristol University Press: Bristol.

Britannica (2022) 'Scientific modeling'. www.britannica.com/science/scientific-modeling (Accessed: 18 August 2022)

Britannica (2023) 'Black Death'. www.britannica.com/event/Black-Death (Accessed: 11 August 2023)

British Medical Association (2022) *BMA Covid Review 4: The Public Health Response by UK Governments to COVID-19*. www.bma.org.uk/media/5980/bma-covid-review-report-4-28-july-2022.pdf (Accessed: 5 February 2024)

Brown, P. and Zinn, J.O. (2022) *Covid-19 and the Sociology of Risk and Uncertainty: Studies of Social Phenomena and Social Theory across 6 Continents*. Springer Nature: Geneva.

Brzezinski, M. (2021) 'The impact of past pandemics on economic and gender inequalities', *medRxiv*, posted 1 May. DOI: https://doi.org/10.1101/2021.04.28.21256239.

Buranyi, S. (2020a) 'The WHO v coronavirus: why it can't handle the pandemic', *Guardian*, 10 April. www.theguardian.com/news/2020/apr/10/world-health-organization-who-v-coronavirus-why-it-cant-handle-pandemic (Accessed: 26 November 2020)

Buranyi, S. (2020b) 'Covid UK: scientists at loggerheads over approach to new restrictions', *Guardian*, 22 September. www.theguardian.com/science/2020/sep/22/scientists-disagree-over-targeted-versus-nationwide-measures-to-tackle-covid (Accessed: 26 July 2023)

Burton, T. (2023) 'AI misinformation attacks are inevitable, warns US expert', *Financial Review*, 23 July. www.afr.com/politics/federal/ai-misinformation-attacks-are-inevitable-warns-us-expert-20230723-p5dqi6 (Accessed: 24 July 2023)

Buus, S. and Olsson, E.-K. (2006) 'The SARS crisis: was anybody responsible?', 14, 2: 71–81.

Cabinet Office (2017) *National Risk Register of Civil Emergencies, 2017 edition*. Cabinet Office: London.

Caduff, C. (2020) 'What went wrong? Corona and the world after the full stop', *Medical Anthropology Quarterly*, 34, 4: 467–487.

Capano, G., Howlett, M., Jarvis, D.S.L., Ramesh, M. and Goyal, N. (2020) 'Mobilizing policy (in)capacity to fight COVID-19: understanding variations in state responses', *Policy and Society*, 39, 3: 285–308. DOI: 10.1080/14494035.2020.1787628

Capoccia, G. and Kelemen, R.D. (2007) 'The study of critical junctures: theory, narrative and counterfactuals in historical institutionalism', *World Politics*, 59, 3: 341–369.

Carlson, C.J., Albery, G.F., Merow, C., Trisos, C.H., Zipfel, C.M., Eskew, E.A., Olival, K.J., Ross, N. and Bansal, S. (2022) 'Climate change increases cross-species viral transmission risk', *Nature*, 607, 7919: 555–562.

Carrington, D. (2020) 'UK strategy to address pandemic threat "not properly implemented"', *Guardian*, 29 March. www.theguardian.com/politics/2020/mar/29/uk-strategy-to-address-pandemic-threat-not-properly-implemented (Accessed: 15 July 2020)

Carstens, A. (2023) Speech to Columbia University, 17 April. Bank for International Settlements. www.bis.org/speeches/sp230417.pdf (Accessed: 19 April 2023)

Casero-Ripollés, A. (2020) 'Impact of Covid-19 on the media system: communicative and democratic consequences of news', *El professional de al information*, 29, 2: e290223, https://doi.org/10.3145/epi.2020.mar.23.

Castel, R. (1991) 'From dangerousness to risk', in G. Burchell, C. Gordon and P. Miller (eds) *The Foucault Effect: Studies in Governmentality*. Harvester Wheatsheaf: London.

Ceci, L. (2023a) 'Most downloaded mobile apps worldwide 2022', 9 January. www.statista.com/statistics/1285960/top-downloaded-mobile-apps-worldwide/ (Accessed: 29 April 2023)

Ceci, L. (2023b) 'TikTok – statistics and facts', 6 April. www.statista.com/topics/6077/tiktok/#topicOverview (Accessed: 29 April 2023)

Center for AI Safety (2023) *Statement on AI Risks*. www.safe.ai/statement-on-ai-risk#open-letter (Accessed: 10 August 2023)

Centers for Disease Control and Prevention (2020) 'Stop the spread of rumours'. www.cdc.gov/coronavirus/2019-ncov/daily-life-coping/share-facts.html (Accessed: 18 July 2020)

Centers for Disease Control and Prevention (2022) 'Nearly one in five American adults who have had COVID-19 still have "Long COVID"', 22 June. www.cdc.gov/nchs/pressroom/nchs_press_releases/2022/20220622.htm (Accessed: 6 February 2024)

Chang, R., Varley, K., Munoz, M., Tam, F. and Makol, M.K. (2021) 'The best and worst places to be as delta wrecks reopening plans', Bloomberg. www.bloomberg.com/graphics/covid-resilience-ranking/ (Accessed: 8 September 2021)

Channel 4 (2023) 'Video: Inside the Shein Machine', First shown 17 October 2022. www.channel4.com/programmes/inside-the-shein-machine-untold (Accessed: 2 August 2023)

Coburn, B.J., Wagner, B.G. and Blower, S. (2009) 'Modeling influenza epidemics and pandemics: insights into the future of swine flu (H1N1)', *BMC Medicine*, 7: 30. DOI: 10.1186/1741-7015-7-30.

Colbourn, T. (2020) 'COVID-19: extending or relaxing distancing control measures', *The Lancet*, 25 March. doi.org/10/1016/S2468-2667(20)30072-4

Connell, R. (2020; orig. 2007) *Southern Theory: The Global Dynamics of Knowledge in Social Science*. Routledge: London and New York.

Council of Europe (2023) 'Covid-19 and human rights', undated. www.coe.int/lv/web/commissioner/thematic-work/covid-19 (Accessed: 10 January 2024)

COVID-19 Genomics UK Consortium (2022) 'Blog/events/women in COG, 18 July'. www.cogconsortium.uk/women-in-cog-in-conversation-with-prof-emma-thomson/ (Accessed: 5 September 2022)

COVID-19 hg (2022) 'About COVID-19 host genetics initiative'. www.covid19hg.org/about/ (Accessed: 5 September 2022)

Credit Suisse Research Institute (2022) *Global Wealth Report 2022: Leading Perspectives to Navigate the Future*. www.credit-suisse.com/media/assets/corporate/docs/about-us/research/publications/global-wealth-report-2022-en.pdf (Accessed: 3 July 2023)

Crow, D. and Waldmeir, P. (2020) 'US anti-lockdown protests: "If you are paranoid about getting sick, don't go out"', *Financial Times*, 22 April. www.ft.com/content/15ca3a5f-bc5c-44a3-99a8-c446f6f6881c (Accessed: 4 July 2020)

Curry, D. (2023) 'Microsoft Teams revenue and usage statistics (2023)', BusinessofApps, Updated 27 April. www.businessofapps.com/data/microsoft-teams-statistics/#:~:text=Microsoft%20Teams%20saw%20a%20huge,Zoom%20from%20February%20to%20June. (Accessed: 1 May 2023)

Cutler, D.M. (2022) 'The economic cost of Long COVID: an update', 16 July. https://scholar.harvard.edu/sites/scholar.harvard.edu/files/cutler/files/long_covid_update_7-22.pdf (Accessed: 7 February 2024)

D'Arpizio, C., Levato, F., Prete, F. and de Montgolfier, J. (2023) 'Renaissance in uncertainty: luxury builds on its rebound', Bain and Company, 17 January. www.bain.com/insights/renaissance-in-uncertainty-luxury-builds-on-its-rebound/ (Accessed: 12 May 2023)

Davis, N. (2022) 'Why did China relax its Covid policy – and should we be worried?', *Guardian*, 30 December. www.theguardian.com/world/2022/dec/29/why-did-china-relax-its-covid-policy-and-should-we-be-worried (Accessed: 8 February 2023)

De Wit, J.B.F., de Ridder, D.T.D., van den Boom, W., Kroese, F.M., van den Putte, B., Stok, F.M., Leurs, M. and de Bruin, M. (2023) 'Understanding public support for COVID-19 pandemic mitigation measures over time: does it wear out?', *Frontiers of Public Health*, 11: 1079992.

Del Boca, D., Oggero, N., Proteta, P. and Rossi, M. (2020) 'Women's and men's work, housework and childcare, before and during COVID-19', *Review of Economics of the Household*, 18: 1001–1017.

Delanty, G. (2021) *Pandemics, Politics, and Society: Critical Perspectives on the Covid-19 Crisis*. De Gruyter: Berlin.

Department of Foreign Affairs and Trade (2023) *The G20*. www.dfat.gov.au/trade/organisations/g20 (Accessed: 8 March 2023)

Department of Health and Aged Care (2022) 'Telehealth hits 100 million services milestone', 17 March. www.health.gov.au/ministers/the-hon-greg-hunt-mp/media/telehealth-hits-100-million-services-milestone#:~:text=Between%2013%20March%202020%20and,services%20to%20support%20their%20patients. (Accessed: 30 May 2023)

Desai, R. (2022) *Capitalism, Coronavirus and War*. Routledge: London.

Despard, M., Bufe, S., Roll, S., Kristansen, K., Fox-Dichter, S.R. and Weiss, M. (2021) 'Employment changes during COVID-19: the importance of employment attachment', Social Policy Institute, Washington University in St. Louis, September 2021. https://bpb-us-w2.wpmucdn.com/sites.wustl.edu/dist/a/2003/files/2021/12/Survey-COVID-Employment_J PMC.pdf (Accessed: 19 April 2023)

Diamond, D. (2022) 'Biden's claim that "pandemic is over" complicates efforts to secure funding', *Washington Post*, 19 September. www.washingtonpost.com/health/2022/09/18/biden-covid-pandemic-over/ (Accessed: 27 September 2022)

Dingwall, R. (2023) 'The sociology of epidemics and pandemics', in A. Petersen (ed) *Handbook on the Sociology of Health and Medicine*. Edward Elgar: Cheltenham and Northampton, MA, pp 455–473.

Dingwall, R., Hoffman, L.M. and Staniland, K. (2013) 'Introduction: why a *Sociology* of Pandemics?', *Sociology of Health and Illness*, 35, 2: 167–173.

Doherty Institute (2021) *Doherty Modelling – Final Report to National Cabinet*, 5 November. www.doherty.edu.au/uploads/content_doc/Synthesis_DohertyModelling_FinalReport__NatCab05Nov.pdf (Accessed: 5 September 2022)

Doherty Institute (2022) 'Modelling'. www.doherty.edu.au/our-work/institute-themes/viral-infectious-diseases/covid-19/covid-19-modelling/modelling (Accessed: 5 September 2022)

Dowell, K.A., Lo Presto, C.T. and Sherman, M.F. (1991) 'When are AIDS patients to blame for their disease? Effects of patients' sexual orientation and mode of transmission', *Psychological Reports*, 69: 211–219.

Drake, J. (2020) 'The science behind London's Christmas coronavirus lockdown', *Forbes*, 19 December. www.forbes.com/sites/johndrake/2020/12/19/the-science-behind-londons-christmas-lockdown/?sh=5be2563d401c (Accessed: 1 September 2022)

Dubos, R. (1996; orig. 1959) *Mirage of Health: Utopias, Progress, and Biological Change*. Rutgers University Press: New Brunswick, NJ.

Duckett, S. (2022) 'Public health management of the COVID-19 pandemic in Australia: the role of the Morrison Government', *International Journal of Environmental Research and Public Health*, 19, 16: 10400.

Dunatchik, A., Gerson, K. and Stritzel, H. (2021) 'Gender, parenting, and the rise of remote work during the pandemic: implications for domestic inequality in the United States', *Gender & Society*, 35, 2: 194–205.

Durkheim, E. (1915) *The Elementary Forms of the Religious Life*. George Allen & Unwin Ltd: London.

Dyer, C. (2020) 'Report of UK's preparedness leaves questions unanswered, says doctor', *The BMJ*, 371: m4499.

Dyvik, E.H. (2024) 'Impact of coronavirus pandemic on the global economy – statistics and facts', *Statista*, 10 January. www.statista.com/topics/6139/covid-19-impact-on-the-global-economy/#topicOverview (Accessed: 23 January 2024)

Englemann, L., Montgomery, C.M., Sturdy, S. and Moreno Lozano, C. (2023) 'Domesticating models: on contingency of Covid-19 modelling in UK media and policy', *Social Studies of Science*, 53, 1: 121–145.

Enserink, M. and Kupferschmidt, K. (2020) 'Mathematics of life and death: how disease models shape national shutdowns and other pandemic policies', *Science*, 25 March. www.science.org/content/article/mathematics-life-and-death-how-disease-models-shape-national-shutdowns-and-other (Accessed: 8 September 2022)

EPIWATCH (2022) 'Research program'. https://kirby.unsw.edu.au/project/epiwatch-prevent-next-pandemic-epidemic-intelligence (Accessed: 12 September 2022)

Eubanks, V. (2018) *Automating Inequality: How High-Tech Tools Profile, Police, and Punish the Poor*. St Martin's Press: New York.

European Institute for Gender Equality (2021) 'Artificial Intelligence, platform work and gender equality'. https://eige.europa.eu/publications-resources/publications/artificial-intelligence-platform-work-and-gender-equality-report (Accessed: 29 May 2023)

European Union (2020a) *EU Disinfo Lab*. www.disinfo.eu/coronavirus (Accessed: 9 June 2020)

European Union (2020b) 'COVID-19 conspiracy theories: comparative trends in Italy, France, and Spain'. www.disinfo.eu/publications/covid-19-conspiracy-theories-comparative-trends-in-italy-france-and-spain (Accessed: 10 June 2020)

Ezrati, M. (2023) 'Covid's long shadow still spreads over commercial real estate', *Forbes*, 19 March. www.forbes.com/sites/miltonezrati/2023/03/17/covids-long-shadow-still-spreads-over-commercial-real-estate/?sh=45fbf14927ae (Accessed: 2 August 2023)

Fandos, N. and Sanger-Katz, M. (2020) 'One group of older Americans is ignoring coronavirus advice: members of congress', *New York Times*, 11 March. www.nytimes.com/2020/03/11/upshot/coronavirus-older-lawmakers-congress-risk.html (Accessed: 2 June 2020)

Fairchild, A.L., Bayer, R., Green, S.H., Colgrove, J., Kilgore, E., Sweeney, M. and Varma, J.K. (2018) 'The two faces of fear: a history of hard-hitting public health campaigns against tobacco and AIDS', *American Journal of Public Health*, 108, 9: 1180–1186.

Fassin, D. and Fourcade, M. (eds) (2021) *Pandemic Exposures: Economy and Society in the Time of Corovirus*. Hau Books: Chicago.

Feierstein, D. (2023) *Social and Political Representations of the COVID-19 Crisis*. Routledge: London.

Ferguson, N.M., Laydon, D., Nedjati-Gilani, G. et al (2020) 'Report 9: impact of non-pharmaceutical interventions (NPIs) to reduce COVID-19 mortality and healthcare demand'. www.imperial.ac.uk/media/imperial-college/medicine/sph/ide/gida-fellowships/Imperial-College-COVID19-NPI-modelling-16-03-2020.pdf (Accessed: 9 September 2022)

Fiske, M., Livingstone, A. and Winona Pit, S. (2020) 'Telehealth in the context of COVID-19: changing perspectives in Australia, the United Kingdom, and the United States', *Journal of Medical Internet Research*, 22, 6: e19264.

Fontanel, J. (2023) 'The war in Ukraine: an unexpected effect of the COVID-19 pandemic', *Visions of Humanity*. www.visionofhumanity.org/the-war-in-ukraine-an-unexpected-effect-of-the-covid-19-pandemic/ (Accessed: 24 March 2023)

Forbes, K. (2022) 'Covid-19 shrank household size, lifted prices', Your Investment Property, 24 August. www.yourinvestmentpropertymag.com.au/expert-insights/kate-forbes/covid-19-shrank-household-size-lifted-prices (Accessed: 16 May 2023)

Frandsen, M. (2020) 'Viral pandemics are the new normal', Thrive Global, 10 April. https://thriveglobal.com/stories/viral-pandemics-the-new-normal-economic-public-health-crisis-coronavirus-pandemic/ (Accessed: 18 December 2022)

Fraser, C., Donnelly, C., Cauchemez, S. et al (2009) 'Pandemic potential of a strain of influenza A (H1N1): early findings', *Science*, 324, 5934: 1557–1561.

Fraser, N., Brierly, L., Dey, G., Polka, J.K., Pálfy, M., Nanni, F. and Coates, J.A. (2021) 'The evolving role of preprints in the dissemination of COVID-19 research and their impact on the science communication landscape', *PLOS Biology*, 22 April. https://journals.plos.org/plosbiology/article?id=10.1371/journal.pbio.3000959 (Accessed: 15 February 2024)

G20 (2024) 'G20 – Background brief', https://www.g20.in/en/docs/2022/ G20_Background_Brief.pdf (Accessed: 12 June 2024)

Gadarian, S.K., Goodman, S.W. and Pepinsky, T.B. (2022) (eds) *Pandemic Politics: The Deadly Toll of Partisanship in the Age of COVID*. Princeton University Press: Princeton, NJ.

Gall Myrick, J. (2015) *The Role of Emotions in Preventive Health Communication*. Lexington Books: Lanham, MD.

Gandenberger, M.K., Knotz, C.M., Fossati, F. and Bonoli, G. (2023) 'Conditional solarity – attitudes towards support for others during the 2020 COVID-19 pandemic', *Journal of Social Policy*, 52, 4: 943–961.

Gao, W. and Wouters, J. (2023) 'The G20's contribution to sustainable development: a perspective from China', in D. Lesage and J. Wouters (eds) *The G20, Development and the UN Agenda 2030*. Routledge: London and New York.

Geddes, L. (2022) 'Next pandemic may come from melting glaciers, new data shows', *Guardian*, 19 October. https://amp.theguardian.com/science/ 2022/oct/19/next-pandemic-may-come-from-melting-glaciers-new-data- shows (Accessed: 7 August 2023)

Geyser, W. (2023) 'How does the TikTok algorithm work?', Influencer Marketing Hub. https://influencermarketinghub.com/tiktok-algorithm/ (Accessed: 5 May 2023)

Ghebreyesus, T.A. and Swaminathan, S. (2021) 'Get ready for AI in pandemic response and healthcare', *BMJ Opinion*, 28 October. https://blogs.bmj. com/bmj/2021/10/28/get-ready-for-ai-in-pandemic-response-and-hea lthcare/ (Accessed: 3 May 2023)

Gieryn, T.F. (1999) *Cultural Boundaries of Science: Credibility on the Line*. University of Chicago Press: Chicago and London.

Gilbert, N. (2022) 'Climate change will force new animal encounters – and boost viral outbreaks', *Nature*, 28 April. www.nature.com/articles/d41 586-022-01198-w (Accessed: 6 July 2023)

Gisaid (2022) 'About Us'. https://gisaid.org/about-us/history/ (Accessed: 5 September 2022)

Glass, J.M. (1997) *'Life Unworthy of Life': Racial Phobia and Mass Murder in Hitler's Germany*. Basic Books: New York.

Gobbi, S., Beata Plomecka, M., Ashraf, Z., Radzinski, P., Neckels, R., Lazzen, S., Dedic, A., Bakalovic, A., Hrustic, L., Skórko, B., Es haghi, S. et al (2020) 'Worsening of preexisting psychiatric conditions during the COVID-19 pandemic', *Frontiers of Psychiatry*, 11: Article 581426. DOI: 10.3389/fpsyt.2020.581426.

Goffman, E. (1986; orig. 1974) *Frame Analysis: An Essay on the Organization of Experience*. Northeastern University Press edition: York.

Gostin, L.O., Moon, S. and Mason Meier, B. (2020) 'Reimagining global health governance in the age of COVID-19', *American Journal of Public Health*, 110, 11: 1615–1619.

Grand View Research (2020) 'Affective computing market size, share and trends analysis report by technology (touch based, touchless), by software, by hardware, and end-user (healthcare automotive), and segment forecasts, 2020–2027'. www.grandviewresearch.com/industry-analysis/affective-computing-market (Accessed: 4 May 2023)

Grand View Research (2023) 'Data centre power market size, share and trends analysis report by product (UPS, PDU, Busway), by end-user (IT & Telecom, BFSI, Retail, Government), by region (Asia Pacific, North America), and segment forecasts, 2023–2030'. www.grandviewresearch.com/industry-analysis/data-center-power-market (Accessed: 26 May 2023)

Gray, M.L. and Siddharth, S. (2019) *Ghost Workers: How to Stop Silicon Valley From Building a New Global Underclass*. Houghton Mifflin Harcourt: Boston, MA and New York.

Great Barrington Declaration (2022a) 'Why was The Declaration written?' https://gbdeclaration.org/why-was-the-declaration-written/ (Accessed: 26 September 2022)

Great Barrington Declaration (2022b) 'The Great Barrington Declaration'. https://gbdeclaration.org/#read (Accessed: 26 September 2022)

Great Barrington Declaration (2022c) 'Signatures' https://gbdeclaration.org/view-signatures/ (Accessed: 26 September 2022)

Green, T. and Fazi, T. (2023) *The Covid Consensus: The Global Assault on Democracy and the Poor – A Critique from the Left*. Hurst Publishers: London.

Grenfell, O. (2023) 'COVID claimed more Australian lives in first three weeks of 2023 than all of 2020', *World Socialist Website*, 22 January. www.wsws.org/en/articles/2023/01/23/ttqb-j23.html (Accessed: 6 February 2023)

Gruszka, K. and Böhm, M. (2022) 'Out of sight, out of mind? (In) visibility of/in platform-mediated work', *New Media and Society*, 24, 8: 1852–1871.

Guglielmi, L. (2020) 'Fast corona tests: what they can and can't do', *Nature*, 585, 7826: 496–498. https://doi.org/10.1038/d41586-020-02661-2.

Gunningham, N., Grabosky, P. and Sinclair, D. (1998) 'Smart regulation: an institutional perspective', *Law and Policy*, 19, 4: 363–414.

Gupta, S., Rouse, B.T. and Sarangi, P.P. (2021) 'Did climate change influence the emergence, transmission and expression of the COVID-19 pandemic?', *Frontiers in Medicine*, 8: 769208. DOI: 10.3389/fmed.2021.769208.

Habermas, J. (1976; orig. 1973) *Legitimation Crisis*. Translated by T. McCarthy. Heinemann: London.

Hacking, I. (1999) *The Social Construction of What?* Harvard University Press: Cambridge, MA.

Haider, N., Rothman-Ostrow, P., Osman, A.Y. et al (2020) 'COVID-19 – Zoonosis or emerging infectious diseases?', *Frontiers in Public Health*, 26 November. https://doi.org/10.3389/fpubh.2020.596944 (Accessed: 6 September 2021)

Hao, K. (2020) 'Nearly half of Twitter accounts pushing to reopen America may be bots', *MIT Technology Review*, 21 May. www.technologyreview.com/2020/05/21/1002105/covid-bot-twitter-accounts-push-to-reopen-america/ (Accessed: 29 June 2020)

Harvey, D. (1990) *The Condition of Postmodernity: An Enquiry into the Origins of Cultural Change*. Blackwell: Cambridge, MA.

Hayden, L., Warren-Norton, K., Chaze, F. and Roberts, R. (2023) 'Pandemic stories: the voices of older adults', *Canadian Journal of Aging*, 42, 1: 154–164.

Hoover, A. (2023) 'The end of the Zoom boom', *Wired*, 8 February. www.wired.com/story/zoom-layoffs-future/ (Accessed: 1 May 2023)

Horton, H. (2023) 'UK is Europe's worst private jet polluter, study finds', *Guardian*, 30 March. www.theguardian.com/environment/2023/mar/30/uk-is-worst-private-jet-polluter-in-europe-study-finds (Accessed: 19 April 2023)

Horton, R. (2020a) 'Offline: COVID-19 is not a pandemic', *Lancet*, 24 September. DOI: 10.1016/S0140-6736(20)32000-6.

Horton, R. (2020b) *The COVID-19 Catastrophe: What's Gone Wrong and How to Stop It Happening Again*. Polity: Cambridge.

House of Commons Science, Innovation and Technology Committee (2023) *Reproducibility and Research Integrity*, Sixth Report of Session 2022–23. https://publications.parliament.uk/pa/cm5803/cmselect/cmsctech/101/report.html (Accessed: 18 May 2023)

Hswen, Y., Xu, X., Hing, A., Hawkins, J.B., Brownstein, J.S. and Gee, G.C. (2021) 'Association of "#covid-19" versus "#chinesevirus" with anti-Asian sentiments on Twitter: March 9–23, 2020', *American Journal of Public Health*, 111, 5: 758–973.

Human Rights Watch (2020) 'Human rights dimensions of COVID-19 response', 19 March. www.hrw.org/news/2020/03/19/human-rights-dimensions-covid-19-response (Accessed: 10 January 2024)

Hutchison, B. (2020) 'Coronavirus lockdowns spark acts of resistance despite warnings of health consequences', *American Broadcasting Commission News*, 22 May. https://abcnews.go.com/US/coronavirus-lockdowns-spark-acts-resistance-warnings-health-consequences/story?id=70650668 (Accessed: 4 July 2020)

Ibbitson, J. and Perkins, T. (2010) 'How Canada made the G20 happen', *The Globe and Mail*, 19 June. www.theglobeandmail.com/news/world/how-canada-made-the-g20-happen/article4322767/?page=all (Accessed: 8 March 2023)

IHME (2021) 'COVID-19 projections'. https://covid19.healthdata.org/uni ted-states-of-america?view=cumulative-deaths&tab=trend (Accessed: 7 July 2021)

Institute for Economics and Peace (2021) *Global Peace Index 2021: Measuring Peace in a Complex World*. Institute for Economics and Peace: Sydney, www. visionofhumanity.org/wp-content/uploads/2021/06/GPI-2021-web-1.pdf (Accessed: 25 January 2024)

Internet Encyclopedia of Philosophy (2022) 'Jeremy Bentham (1748–1832)'. https://iep.utm.edu/jeremy-bentham/ (Accessed: 16 September 2022)

Interpol (2020) *Threats and Trends Child Sexual Exploitation and Abuse*, Covid-19 Impact. September. https://www.interpol.int/en/News-and-Events/ News/2020/INTERPOL-report-highlights-impact-of-COVID-19-on-child-sexual-abuse (Accessed: 1 July 2024)

Iqbal, M. (2023) 'Zoom revenue and usage statistics (2023)', Updated 26 April. www.businessofapps.com/data/zoom-statistics/ (Accessed: 1 May 2023)

Ireson, J., Taylor, A., Richardson, E., Greenfield, B. and Jones, G. (2022) 'Exploring invisibility and epistemic injustice in Long Covid – A citizen science qualitative analysis of patient stories from an online Covid community', *Health Expectations*, 25: 1753–1765.

Jonas, O.B. (2013) *Pandemic Risk*. Background Paper. The World Bank: Geneva. https://www.worldbank.org/content/dam/Worldbank/document/HDN/ Health/WDR14_bp_Pandemic_Risk_Jonas.pdf (Accessed: 17 July 2024)

Jordà, Ò. and Nechio, F. (2023) 'Inflation and wage growth since the pandemic', *European and Economic Review*, July, 156: 104474, DOI: 10.1016/ j.euroecorev.2023.104474.

Jungkunz, S. (2021) 'Political polarization during the COVID-19 pandemic', *Frontiers of Political Science*, 3, https://doi.org/10.3389/fpos.2021.622512.

Kartono, R., Salahudin, N.A. and Sihidi, I.T. (2022) 'Covid-19 stigmatization: a systematic literature review', *Journal of Public Health Research*, 11, 3: 22799036221115780.

Katz, I.T., Weintraub, R., Bekker, L-G. and Brandt, A.M. (2021) 'From vaccine nationalism to vaccine equity – finding a path forward', *New England Journal of Medicine*, 384: 1281–1283, 8 April. www.nejm.org/doi/ full/10.1056/NEJMp2103614 (Accessed: 4 May 2021)

Kelly, A. and Pattison, P. (2021) '"A pandemic of abuses": human rights under attack during Covid, says UN head', *Guardian*, 22 February. www.theguard ian.com/global-development/2021/feb/22/human-rights-in-the-time-of-covid-a-pandemic-of-abuses-says-un-head (Accessed: 10 January 2024)

Kelly, H. (2011) 'The classical definition of a pandemic is not elusive', *Bulletin of the World Health Organization*, 89, 7: 540–541. https://apps.who.int/iris/ handle/10665/270942/ (Accessed: 10 August 2023)

Kelly, M. (2023) 'Strongest private jet sales ever reported by IADA members', *Corporate Jet Investor*, 19 January. www.corporatejetinvestor.com/news/strongest-private-jet-sales-ever-reported-by-iada-members/ (Accessed: 19 April 2023)

Kermack, W.O. and McKendrick, A.G. (2017; orig. 1927) 'A seminal contribution to the mathematical theory of epidemics (1927)', in I.M. Foppa (ed) *A Historical Introduction to Mathematical Modeling of Infectious Diseases*. Academic Press: Cambridge, MA.

Kieslich, K., Fiske, A., Gaille, M. et al (2023) 'Solidarity during the COVID-19 pandemic: evidence from a nine-country interview study of Europe', *Medical Humanities*, 49: 511–520.

Kim, J. (2023) 'Data brokers and the sale of Americans' mental health data', Cyber Policy Program, February. https://techpolicy.sanford.duke.edu/wp-content/uploads/sites/4/2023/02/Kim-2023-Data-Brokers-and-the-Sale-of-Americans-Mental-Health-Data.pdf (Accessed: 29 January 2024)

Kiran, H. and Defensor, G. (2023) '10 data broker statistics you need to know', *TechJury* blog, 6 June. https://techjury.net/blog/data-broker-statistics/ (Accessed: 24 July 2023)

Kirchgaessner, S. (2020) 'Mobile phone industry explores worldwide tracking of users', *Guardian*, 25 March. www.theguardian.com/world/2020/mar/25/mobile-phone-industry-explores-worldwide-tracking-of-users-coronavirus (Accessed: 22 August 2022)

Kirkland, T. and Fang, G. (eds) (2023) *Pandemedia: How COVID Changed Journalism*. Monash University Publishing: Clayton.

Klein, N. (2008) *The Shock Doctrine: The Rise of Disaster Capitalism*. Penguin: New York

Klein, N. (2020) 'Coronavirus capitalism – and how to beat it', *The Intercept*, 17 March. https://theintercept.com/2020/03/16/coronavirus-capitalism/ (Accessed: 28 May 2020)

Knight, D. (2021) 'COVID-19 pandemic origins: bioweapons and the history of laboratory leaks', *Southern Medical Journal*, 114, 8: 565–467. www.ncbi.nlm.nih.gov/pmc/articles/PMC8300139/ (Accessed: 8 September 2021)

Koffman, J., Gross, J., Edkind, S.N. and Selman, L. (2020) 'Uncertainty and COVID-19: how are we to respond?', *Journal of the Royal Society of Medicine*, 113, 6: 211–216.

Komaroff, A.L. and Lipkin, W.I. (2023) 'ME/CFS and Long COVID share similar symptoms and biological abnormalities: road map to the literature', *Frontiers of Medicine*, 10, https://doi.org/10.3389/fmed.2023.1187163.

Kucharski, A. (2021) 'How modelling Covid has changed the way we think about epidemics', *Guardian*, 4 January. www.theguardian.com/commentisfree/2021/jan/04/covid-model-epidemic-collaboration-experiment (Accessed: 22 August 2022)

Last, J.M. (ed) (2001) *A Dictionary of Epidemiology*. 4th edn. New York: Oxford University Press.

Latour, B. (2021) *After Lockdown: A Metamorphisis*. Polity: Cambridge.

Latour, B. and Woolgar, S. (1986; orig. 1979) *Laboratory Life: The Construction of Scientific Facts*. Princeton University Press: Princeton, NJ.

Law, J. (2004) *After Method: Mess in Social Science Research*. Routledge: London.

Lawford-Smith, H. (2022) 'Was lockdown life worth living?', *Monash Bioethics Review*, 40: 40–61.

Le Grand, C. (2022) *Lockdown*. Monash University Press: Clayton, Victoria.

Leach, M., MacGregor, H., Ripoll, S., Scoones, I. and Wilkinson, A. (2022) 'Rethinking disease preparedness: incertitude and the politics of knowledge', *Critical Public Health*, 32, 1: 82–96.

Lemieux, A., Colby, G.A., Poulain, A.J. and Aris-Brosou, S. (2022) 'Viral spillover risk increases with climate change in High Arctic lake sediments', *Proceedings of the Royal Society B.*, 289: 20221073. https://royalsocietypublishing.org/doi/epdf/10.1098/rspb.2022.1073 (Accessed: 7 August 2023)

Levine, M. (2020) 'Governors were warned of a pandemic years ago, told to stockpile. Why didn't they do more?', *ABC News*, 29 April. https://abcnews.go.com/Politics/governors-warned-pandemic-years-ago-told-stockpile-didnt/story?id=70331277 (Accessed: 14 July 2020)

Lewis, C., Phillipson, C., Yarker, S. and Lang, L. (2023) *COVID-19, Inequality and Older People: Everyday Life During the Pandemic*. Bristol University Press: Bristol.

Lewis, W. (2020) 'Disaster response expert explains why the U.S. wasn't more prepared for the pandemic', *USC Dornsife*, 24 March. https://dornsife.usc.edu/news/stories/why-u-s-wasnt-better-prepared-for-the-coronavirus/ (Accessed: 11 August 2023)

Liedtke, M. (2023) 'Google unveils AI-powered search engine in tech race', *Financial Review*, 11 May. www.afr.com/technology/google-unveils-ai-powered-search-engine-in-tech-race-20230511-p5d7in (Accessed: 15 May 2023)

Lockdown Sceptics (2020) 'Latest news'. https://lockdownsceptics.org/ (Accessed: 4 July 2020)

Long, H., Dam, A.V., Fowers, A. and Shapiro, L. (2020) 'The covid-19 recession is the most unequal in modern U.S. history', *Washington Post*, 30 September. www.washingtonpost.com/graphics/2020/business/coronavirus-recession-equality/?itid=lk_inline_manual_13 (Accessed: 26 April 2021)

Lorenz, T. (2023) 'An influencer's AI clone will be your girlfriend for $1 a minute', *Washington Post*, 13 May. www.washingtonpost.com/technology/2023/05/13/caryn-ai-technology-gpt-4/ (Accessed: 15 May 2023)

Lourenco, P.R., Kaur, G., Alison, M. and Evetts, T. (2021) 'Data sharing and collaborations with Telco data during the COVID-19 pandemic: a Vodaphone case study', *Data Policy*, 3: e33, DOI: 10.1017/dap.2021.26.

Lupton, D. (2022) *COVID Societies: Theorising the Coronavirus Crisis.* Routledge: London.

Luu, T. (2022) 'Reduced cancer screening due to lockdowns of the COVID-19 pandemic: reviewing impacts and ways to counteract the impacts', *Frontiers of Oncology*, 12: 955377.

Lyon, D. (2022) *Pandemic Surveillance.* Polity Press: Cambridge.

Madgavkar, A., White, O., Krishnan, M., Mahajan, D. and Azcue, X. (2020) 'COVID-19 and gender enquality: countering the regressive effects', 15 July. www.mckinsey.com/featured-insights/future-of-work/covid-19-and-gender-equality-countering-the-regressive-effects (Accessed: 18 May 2023)

Maes, C. and Vandenbosch, L. (2022) 'Adolescent girl's Instagram and TikTok use: examining relations with body image-related constructs over time using random intercept cross-lagged panel models', *Body Image*, 41: 453–459.

Magal, P. and Ruan, S. (2014) 'Susceptible-infectious-recovered models revisited: from the individual level to the population level', *Mathematical Biosciences*, 250: 26–40.

Maharasigam-Shah, E. and Vaux, P. (2021) *'Climate lockdown' and the Culture Wars: How COVID-19 Sparked a New Narrative against Climate Action.* Institute for Strategic Dialogue: London.

Malnick, E. (2020) 'Coronavirus advice being ignored by almost a quarter of 25 to 34 year-olds, poll finds', *Telegraph*, 21 March. www.telegraph.co.uk/news/2020/03/21/coronavirus-advice-ignored-almost-quarter-25-34-year-olds-poll/ (Accessed: 2 June 2020)

Mannix, L. (2021) 'The end of the pandemic? More like the rebirth of the pandemic age', *Sydney Morning Herald*, 13 March. www.smh.com.au/national/the-end-of-the-pandemic-more-like-the-rebirth-of-the-pandemic-age-20210311-p579wo.html (Accessed: 20 April 2021)

Marcovitch, H. (ed) (2017) *Black's Medical Dictionary.* 43rd edn. Bloomsbury Publishing: London and New York.

Marjanovic, S., Romanelli, R.J., Ali, G-C., Leach, B., Bonsu, M., Rodriguez-Rincon, D. and Ling, T. (2022) 'COVID-19 Genomics UK (COG-UK) Consortium: Final Report', *Rand Health Quarterly*, 9, 4: 24. PMCID: PMC9519096.

Mason, R. (2020) 'Boris Johnson boasted of shaking hands on day Sage warned not to', *Guardian*, 5 May. www.theguardian.com/politics/2020/may/05/boris-johnson-boasted-of-shaking-hands-on-day-sage-warned-not-to (Accessed: 2 June 2020)

Maxmen, A. and Mallapaty, S. (2021) 'The COVID lab-leak hypothesis: what scientists do and don't know', *Nature*, 8 June. www.nature.com/articles/d41586-021-01529-3 (Accessed: 7 September 2021)

McBryde, E.S., Meehan, M.T., Adegboye, O.A., Adekunle, A.I., Caldwell, J.M., Pak, A., Rojas, D.P., Williams, B.M. and Trauer, J.M. (2020) 'Role of modelling in COVID-19 policy development', *Paediatric Respiratory Reviews*, 35: 57–60.

McClellan, M., Udayakumar, K., Merson, M. and Edson, G. (2021) 'Reducing Global COVID Vaccine Shortages: New Research and Recommendations for US Leadership', 15 April, Duke Margolis Center for Health Policy, Duke Global Health Innovation Center, and Duke Global Health Institute. https://healthpolicy.duke.edu/sites/default/files/2021-04/US%20Leadership%20for%20Global%20Vaccines_1.pdf (Accessed: 10 August 2023)

McKinsey and Company (2020) 'How COVID-19 has pushed companies over the technology tipping point – and transformed business forever', 5 October. www.mckinsey.com/capabilities/strategy-and-corporate-fina nce/our-insights/how-covid-19-has-pushed-companies-over-the-technol ogy-tipping-point-and-transformed-business-forever#/ (Accessed: 24 April 2023)

Mehta, R., Thakur, S. and Chakraborty, D. (2023) *Pandemic of Perspectives: Creative Re-Imaginings*. Routledge: London.

Melendez, P. (2020) '"This is a war": Cuomo pleads for help from doctors across U.S. as coronavirus death toll surges', *Daily Beast*, 30 March. www. thedailybeast.com/andrew-cuomo-pleads-for-help-from-doctors-nurses-across-us-amid-coronavirus-pandemic (Accessed: 14 July 2020)

Melendez, S. and Pasternack, A. (2019) 'Here are the data brokers quietly buying and selling your personal information', *Fast Company*, 2 March. www.fastcompany.com/90310803/here-are-the-data-brokers-quietly-buy ing-and-selling-your-personal-information (Accessed: 29 January 2024)

Meltzer, M.I., Santibanez, S., Fischer, L.S. et al (2016) 'Modeling in real time during the Ebola response', Supplement, *Morbidity and Mortality Weekly Report*, 65, 3: 85–89.

Merrick, R. (2020) 'Dominic Cumming's flouting of lockdown rules "clearly undermined" fight against coronavirus, ex-civil service chief admits', *Independent*, 21 October. www.independent.co.uk/news/uk/ politics/dominic-cummings-lockdown-rules-break-drive-car-durham-coronavirus-civil-service-boris-johnson-b1200265.html (Accessed: 3 December 2020)

Mertens, G., Engelhard, I.M., Novacek, D.M. and McNally, R.J. (2023) 'Managing fear during pandemics: risks and opportunities', *Perspectives on Psychological Science*, 26 June: 17456916231179720. DOI: 10.1177/ 17456916231178720.

Meskó, B. (2022) 'COVID-19's impact on digital health adoption: the growing gap between technological and cultural transformation', *Journal of Medical Internet Research Human Factors*, 9, 3: e3826.

Metzler, H., Pellert, M. and Garcia, D. (2022) 'Chapter 4. Using social media data to capture emotions before and during COVID-19'. https://happiness-report.s3.amazonaws.com/2022/WHR+22_Ch4.pdf (Accessed: 5 May 2023)

Mijatović, D. (2020) 'Human Rights Talk: Covid-19 and Human Rights – Lessons learned from the pandemic', 10 December. https://rm.coe.int/human-rights-rights-talk-covid-19-and-human-rights-lessons-learned-fro/1680a0a7c3 (Accessed: 10 January 2024)

Miller, M.B. and Riechert, B.P. (2000) 'Interest group strategies and journalist norms: news media framing of environmental issues', in S. Allan, B. Adam and C. Carter (eds) *Environmental Risks and the Media*. Routledge: London and New York.

Mol, A. (2002) *The Body Multiple: Ontology in Medical Practice*. Duke University Press: Durham, NC.

Moloney, K. and Moloney, S. (2020) 'Australian quarantine policy: from centralization to coordination with mid-pandemic COVID-19 shifts', *Public Administration Review*, 80, 4: 671–682.

Monaghesh, E. and Hajizadeh, A. (2020) 'The role of telehealth during COVID-19 outbreak: a systematic review based on current evidence', *BMC Public Health*, 20, Article: 1193. https://doi.org/10.1186/s12889-020-09301-4 (Accessed: 11 August 2023)

Moyo, I. and Ndlovu-Gatsheni, S.J. (eds) (2024) *The COVID-19 Pandemic and the Politics of Life*. Routledge: London.

Muggah, R. and Florida, R. (2020) 'Megacity slums are incubators of disease – but coronavirus response isn't helping the billion people who live in them', *The Conversation*, 15 May, https://theconversation.com/megacity-slums-are-incubators-of-disease-but-coronavirus-response-isnt-helping-the-billion-people-who-live-in-them-138092?utm_medium=email&utm_campaign=Latest%20from%20The%20Conversation%20for%20May%2018%202020%20-%201624615593&utm_content=Latest%20from%20The%20Conversation%20for%20May%2018%202020%20-%201624615593+CID_85e0aca596e1c15f5ac3fc9b7c62a62e&utm_source=campaign_monitor_global&utm_term=Megacity%20slums%20are%20incubators%20of%20disease%20%20but%20coronavirus%20response%20isnt%20helping%20the%20billion%20people%20who%20live%20in%20them (Accessed: 8 August 2023)

Munir, F. (2021) 'Mitigating COVID: impact of COVID-19 lockdown and school closure on children's wellbeing', *Social Sciences*, 10, 387, https://doi.org/10.3390/socsci10100387.

Nair, A. (2020) 'Coronavirus: warnings issued as thousands "recklessly" ignore social distancing advice', *Sky News*, 23 March. https://news.sky.com/story/coronavirus-warnings-issued-as-thousands-recklessly-ignore-social-distancing-advice-11962127 (Accessed: 2 June 2020)

National Cancer Institute (2022) 'Working to close the cancer screening gap caused by COVID', 17 May. www.cancer.gov/news-events/cancer-currents-blog/2022/covid-increasing-cancer-screening (Accessed: 2 February 2024)

National Institutes of Health (2023) 'Risky alcohol use: an epidemic inside the COVID-19 pandemic', 28 July. https://covid19.nih.gov/news-and-stories/risky-drinking-alcohol-use-epidemic-inside-covid-19-pandemic (Accessed: 2 February 2024)

Navarro, V. (2020) 'The consequences of neoliberalism in the current pandemic', *International Journal of Health Services*, 50, 3: 271–275.

Navuluri, N., Solomon, H.S., Hargett, C.W. and Kussin, P.S. (2021) 'Where have all the heroes gone?', *Medical Anthropology*, 40, 3: 209–213.

Newman, K. (2020) 'Survey: amid the COVID-19 pandemic, urbanites are eyeing the suburbs', US News, 6 May. www.usnews.com/news/cities/articles/2020-05-06/a-recent-survey-suggests-the-pandemic-has-urbanites-eyeing-the-suburbs (Accessed: 15 May 2023)

News.com.au (2020) 'Coronavirus: Sweden's herd immunity approach backfires with low antibody rate', News.com.au, 23 June. www.news.com.au/lifestyle/health/health-problems/coronavirus-swedens-herd-immunity-approach-backfires-with-low-antibody-rate/news-story/5417a20f4f3298d981968d897965c403 (Accessed: 19 July 2023)

Nextstrain (2022) 'Nextstrain: real-time tracking of pathogen evolution'. https://nextstrain.org/ (Accessed: 1 September 2022)

Nie, J.-B., Gilbertson, A.L., de Roubaix, M., Staunton, C., van Niekerk, A., Tucker, J.D. and Rennie, S. (2016) 'Healing without waging war: beyond military metaphors in medicine and HIV cure research', *American Journal of Bioethics*, 16, 10: 3–11. DOI: 10.1080/15265161.2016.1214305.

Nisbet, M.C. and Mooney, C. (2007) 'Framing science', *Science*, 316, 5821: 56.

Noble, S.U. (2018) *Algorithms of Oppression: How Search Engines Reinforce Racism*. New York University Press: New York.

Norberg, J. (2011) 'Arendt in crisis: political thought in between past and future', *College Literature*, 38, 1: 131–149.

O'Malley, P. (2004) *Risk, Uncertainty and Government*. The GlassHouse Press: London.

O'Malley, P. (2009) 'Responsibilization', in A. Wakefield and J. Flemming (eds) *The Sage Dictionary of Policing*. Sage: London and Thousand Oaks, CA.

Oberkampf, W.L., DeLand, S.M., Rutherford, B.M., Diegert, K.V. and Alvin, K.F. (2002) 'Error and uncertainty in modeling and simulation', *Reliability Engineering and System Safety*, 75: 333–357.

OECD (2020a) 'Tracking and tracing COVID: protecting privacy and data while using apps and biometrics', updated 23 April. www.oecd.org/coronavirus/policy-responses/tracking-and-tracing-covid-protecting-privacy-and-data-while-using-apps-and-biometrics-8f394636/ (Accessed: 10 May 2023)

OECD (2020b) 'The impact of the COVID-19 pandemic on jobs and incomes in G20 economies', ILO–OECD paper prepared at the request of G20 Leaders Saudi Arabia's G20 Presidency 2020. www.ilo.org/wcm sp5/groups/public/---dgreports/---cabinet/documents/publication/wcms _756331.pdf (Accessed: 2 February 2024)

OECD (2022) 'First lessons from government evaluations of COVID-19 responses: a synthesis', 21 January. www.oecd.org/coronavirus/policy-responses/first-lessons-from-government-evaluations-of-covid-19-respon ses-a-synthesis-483507d6/ (Accessed: 2 February 2024)

Office of the Australian Information Commissioner (2023) 'Notifiable data breaches report: July to December 2022', 1 March. www.oaic.gov.au/priv acy/notifiable-data-breaches/notifiable-data-breaches-publications/notifia ble-data-breaches-report-july-to-december-2022 (Accessed: 22 May 2023)

Oi-Yee Li, H., Bailey, A., Huynh, D. and Chan, J. (2020) 'YouTube as a source of information on COVID-19: a pandemic of misinformation?', *BMJ Global Health*, 5: e.002604. doi.10.1136/bmjgh-2020-002604.

Olick, R.R., Shaw, J. and Yang, Y.T. (2021) 'Ethical issues in mandating COVID-19 vaccination for health care personnel', *Mayo Clinic Proceedings*, 96, 12: 2958–2962.

Oppenheimer, G. (1995) 'Comment: epidemiology and the liberal arts – toward a new paradigm?', *The American Journal of Public Health*, 85, 7: 918–920.

Our World in Data (2023) 'Number of parameters in notable artificial intelligence systems'. https://ourworldindata.org/grapher/artificial-intellige nce-parameter-count#:~:text=Parameters%20are%20variables%20in%20 an,in%20an%20artificial%20neural%20network (Accessed: 18 May 2023)

Oxfam (2023) *Survival of the Richest: How We Must Tax the Super-Rich Now to Fight Inequality*. https://oi-files-d8-prod.s3.eu-west-2.amazonaws.com/ s3fs-public/2023-01/Survival%20of%20the%20Richest%20Full%20Rep ort%20-English.pdf (Accessed: 17 April 2023)

Pai, M. (2020) 'Covidization of research: what are the risks?', *Nature Medicine*, 26, 1159. DOI: https://doi.org/10.1038/s41591-020-1015-0.

Panovska-Griffiths, J. (2020) 'Can mathematical modelling solve the current COVID-19 crisis?', *BMC Public Health*, 20: 551. https://doi.org/10.1186/ s12889-020-08671-z.

Park, S., Fisher, C., Lee, J.Y. and McGuinness, K. (2020) COVID-19: Australian news and misinformation, 7 July, Available at: https://apo. org.au/node/306728 (Accessed: 28 April 2021)

Parliament of Australia (2020) 'Chapter 4: Cybercrime and cyber-enabled crime', Parliament Joint Committee on Law Enforcement, *Inquiry Into COVID-19, Criminal Activity and Law Enforcement*. www.aph.gov.au/Par liamentary_Business/Committees/Joint/Law_Enforcement/COVID-19/ Report/section?id=committees%2Freportjnt%2F024564%2F73893#footn ote26target (Accessed: 29 January 2024)

Parmet, W.E., Goodman, R. and Farber, A. (2005) 'Individual rights versus the public's health – 100 years after Jacobson v. Massachusetts', *The New England Journal of Medicine*, 352, 7: 652–654.

Parthasarathy, V. (2023) 'Zoom's AI innovations empower people', Zoom Blog, 24 February. https://blog.zoom.us/ai-driven-innovations/ (Accessed: 2 May 2023)

Patterson, S. (2023) *Chaos Kings: How Wall Street Traders Make Billions in the New Age of Crisis*. Scribe Publications: New York.

Peacock, S. (2021) 'Q7A with Sharon Peacock, coronavirus variant hunter', *The Conversation*, 12 February. https://theconversation.com/qanda-with-sharon-peacock-coronavirus-variant-hunter-154808 (Accessed: 5 September 2022)

Peckham, R. (2020) 'COVID-19 and the anti-lessons of history', *The Lancet*, 20 March. www.thelancet.com/journals/lancet/article/PIIS0140-6736(20)30468-2/fulltext (Accessed: 29 April 2021)

Pennington, H. (2022) *COVID-19: The Post-Genomic Pandemic*. Polity Press: Cambridge.

Petersen, A. (2002) 'Replicating our bodies, losing our selves', *Body & Society*, 8, 4: 71–90.

Petersen, A. (2015) *Hope in Health: The Socio-Politics of Optimism*. Palgrave Macmillan: Baskingstoke.

Petersen, A. (2019) *Digital Health and Technological Promise: A Sociological Inquiry*. Routledge: London and New York.

Petersen, A. (2023) *Emotions Online: Feelings and Affordances of Digital Media*. Routledge: London and New York.

Petersen, A. and Lupton, D. (1996) *The New Public Health: Health and Self in the Age of Risk*. Sage: London.

Petersen, A. and Pienaar, K. (2024) 'Competing realities, uncertain diagnoses of infectious disease: mass self-testing for COVID-19 and liminal bio-citizenship', *Sociology of Health and Illness*, 46(S1): 242–260. doi: 10.1111/1467-9566.13694.

Petersen, A., Schermuly, A. and Anderson, A. (2019) 'The shifting politics of patient activism: from bio-sociality to bio-digital citizenship', *Health*, 23, 4: 478–494.

Pfeffer, F.T., Danziger, S. and Schoeni, R.F. (2013) 'Wealth disparities before and after the Great Recession', *The ANNALS of the American Academy of Political and Social Science*, 650, 1: 98–123.

Piquero, A.R., Jennings, W.G., Jemison, E., Kaukinen, C. and Knaul, F.M. (2021) 'Domestic violence during the COVID-19 pandemic: evidence from a systematic review and meta-analysis', *Journal of Criminal Justice*, 74: 101806.

Pitron, G. (2023) *The Dark Cloud: How the Digital World is Costing the Earth*. Scribe Publications: Brunswick and London.

Poirier, A.C., Riaño Moreno, R.D., Takaindisa, L. et al (2023) VIDIIA Hunter; a low-cost, smartphone connected, artificial intelligence-assisted COVID-19 rapid diagnostic platform approved for medical use in the UK', *Frontiers in Molecular Biosciences*, 10. https://doi/org/10.3389/fmolb.2023.1144001.

Raffle, A.E. and Gill, M. (2021) 'Mass screening for asymptomatic SARS-CoV-2', *The BMJ*, 273, n1058. https://doi.org/10.1136/bmj.n1058.

Rajmil, L., Hjern, A., Boran, P., Gunnlaugsson, G. and Kraus de Carmargo, O. (2021) 'Impact of lockdown and school closure on children's health and wellbeing during the first wave of COVID-19: a narrative review', *BMJ Paediatrics Open*, 5, 1: e001043.

Rawlinson, K. (2020) ' "This enemy can be deadly": Boris Johnson invokes wartime language', *Guardian*, 18 March. https://www/theguardian.com/world/2020/mar/17/enemy-deadly-boris-johnson-invokes-wartime-language-coronavirus (Accessed: 15 July 2020)

Read, M. (2023) 'JobKeeper and "excessive" stimulus to blame for high inflation', *Australian Financial Review*, 16 April, p 2.

Read, M. and McIllroy, T. (2023) 'Australia's pandemic population shuffle revealed', *Financial Review*, 7 May. www.afr.com/policy/economy/australia-s-pandemic-population-shuffle-revealed-20230425-p5d317 (Accessed: 11 May 2023)

Reddy, W. (2001) *The Navigation of Feeling: A Framework for the History of Emotions*. Cambridge University Press: Cambridge.

Renn, O. (2008a) *Risk Governance: Coping with Uncertainty in a Complex World*. Earthscan: London.

Renn, O. (2008b) 'Concepts of risk: an interdisciplinary review', *GAIA*, 17, 1: 50–66.

Reserve Bank of Australia (2023) 'A new measure of average household size', 16 March. www.rba.gov.au/publications/bulletin/2023/mar/a-new-measure-of-average-household-size.html (Accessed: 16 May 2023)

Rettig, R.A., Jacobson, P.D., Farquhar, C. and Aubury, W.M. (2007) *False Hope: Bone Marrow Transplantation for Breast Cancer*. Oxford University Press: Oxford.

Reuters (2020) 'Escape from the city? Londoners lead Europe in COVID-inspired dreams of flight', 19 November. www.reuters.com/article/europe-cities-coronavirus-idINL8N2I41HK (Accessed: 15 May 2023)

Reynolds, K. (2021) 'COVID-19 increased the use of AI. Here's why it's here to stay', World Economic Forum, 24 February. www.weforum.org/agenda/2021/02/covid-19-increased-use-of-ai-here-s-why-its-here-to-stay/ (Accessed: 18 May 2023)

Rhodes, T. and Lancaster, K. (2020) 'Mathematical models as public troubles in COVID-19 infection control: following the numbers', *Health Sociology Review*, 29, 2: 177–194.

Rhodes, T. and Lancaster, K. (2022) 'Making pandemics big: on the situational performance of Covid-19 mathematical models', *Social Science & Medicine*, 301: 114907.

Rhodes, T., Lancaster, K. and Rosengarten, M. (2020) 'A model society: maths, models and expertise in viral outbreaks', *Critical Public Health*, 30, 3: 253–256.

Roberts, J.D. and Tehrani, S. (2020) 'Environments, behaviours, and inequalities: reflecting on the impacts of the Influenza and Coronavirus Pandemics in the United States', *Environmental Research and Public Health*, 17, 4484, DOI: 10.3390/ijerph17124484.

Robertson, S. (2021) ' "Herd immunity" not responsible for Sweden's control of COVID-19, say researchers', *News Medical Life Sciences*, 14 July. www.news-medical.net/news/20210714/e2809cHerd-immunitye2809d-not-responsible-for-Swedene28099s-control-of-COVID-19-say-researchers.aspx (Accessed: 19 July 2023)

Rosanwo, D. (2021) 'Cities are not going away according to urban dwellers', The Harris Poll, 2 February. https://theharrispoll.com/briefs/cities-are-not-going-away-according-to-urban-dwellers/ (Accessed: 11 May 2023)

Roy, M., Moreau, N., Rousseau, C., Mercier, A., Wilson, A. and Atlani-Duault, L. (2019) 'Ebola and localised blame on social media: analysis of Twitter and Facebook conversations during the 2014–2015 Ebola Epidemic', *Culture, Medicine, and Psychiatry*, 44, 1: 56–79.

Royal Commission into Aged Care Quality and Safety (2020) *Aged Care and COVID-19: A Special Report*. www.royalcommission.gov.au/system/files/2021-03/aged-care-and-covid-19-a-special-report.pdf (Accessed: 26 January 2024)

Ryan, J.M. (2021) *COVID-19: Volume 1 and 2*. Routledge: London.

Salter, M. and Wong, W.K.T. (2021) *The Impact of COVID-19 on the Risk of Online Child Sexual Exploitation and the Implications for Child Protection and Policing*. Research Report, May. www.end-violence.org/sites/default/files/paragraphs/download/esafety%20OCSE%20report%20-%20salter%20and%20wong.pdf (Accessed: 8 September 2021)

Sample, I. (2023) 'New data links Covid-19's origins to raccoon dogs in Wuhan market', *Guardian*, 18 March. www.theguardian.com/society/2023/mar/17/covid-19-origins-raccoon-dogs-wuhan-market-data?CMP=Share_iOSApp_Other (Accessed: 20 March 2023)

Savulescu, J., Persson, I. and Wilkinson, D. (2020) 'Utilitarianism and the pandemic', *Bioethics*, 34, 6: 620–632.

Scally, G., Jacobson, B. and Abbasi, K. (2020) 'The UK's public health response to covid-19', *The BMJ*, 369:m1932.

Schleicher, A. (2020) *The Impact of COVID-19 on Education: Insights from Education at a Glance 2020*. OECD. www.oecd.org/education/the-impact-of-covid-19-on-education-insights-education-at-a-glance-2020.pdf (Accessed: 11 May 2023)

Schnell, L. (2020) 'Crowded bars and theme parks: why won't some people practice social distancing during coronavirus outbreak?, *USA Today*, 16 March. www.usatoday.com/story/news/nation/2020/03/16/coronavirus-social-distancing-why-people-wont-avoid-each-other/5065228002/ (Accessed: 2 June 2020)

Science Media Centre (2020) 'Expert reaction to letter sent from Donald Trump to Dr Tedros Adhanom, Director-General of the WHO', 19 May. www.sciencemediacentre.org/expert-reaction-to-letter-sent-from-donald-trump-to-dr-tedros-adhanom-director-general-of-the-who/ (Accessed: 25 July 2023)

Shackle, S. (2021) 'Among the Covid sceptics: "We are being manipulated, without a shadow of doubt"', *Guardian*, 8 April. www.theguardian.com/news/2021/apr/08/among-covid-sceptics-we-are-being-manipulated-anti-lockdown (Accessed: 9 February 2024)

Sharma, A., Virami, T., Pathak, V., Sharma, A., Pathak, K., Kumar, G. and Pathak, D. (2022) 'Artificial intelligence-based data-driven strategy to accelerate research, development, and clinical trials of COVID vaccine', *Biomedical Research International*, 2022, Article ID 7205241.

Shorrocks, A., Davies, J. and Lluberas, R. (2022) 'Global wealth distribution 2021', in Credit Suisse Research Institute, *Global Wealth Report 2022: Leading Perspectives to Navigate the Future*. www.credit-suisse.com/media/assets/corporate/docs/about-us/research/publications/global-wealth-report-2022-en.pdf (Accessed: 3 July 2023)

Similarweb (2023) 'Zoom.us'. www.similarweb.com/website/zoom.us/#overview (Accessed: 1 May 2023)

Sirleaf, E.J. and Clark, H. (2021) *Losing Time: End this Pandemic and Secure the Future*. https://theindependentpanel.org/wp-content/uploads/2021/11/COVID-19-Losing-Time_Final.pdf (Accessed: 11 August 2023)

Social Policy Institute (2021) 'New longitudinal socioeconomic impacts of COVID-19 survey calls for sustained public benefit support', 15 September. https://socialpolicyinstitute.wustl.edu/new-longitudinal-socioeconomic-impacts-of-covid-19-survey-calls-for-sustained-public-benefit-support/ (Accessed: 19 April 2023)

Sofonea, M.T., Chauchemez, S. and Boelle, P.-Y. (2022) 'Epidemic models: why and how to use them', *Anaesthesia Critical Care & Pain Medicine*, 41: 101048.

Sol Hart, P., Chinn, S. and Soroka, S. (2020) 'Politicization and polarization in COVID-19 news coverage', *Science Communication*, 42, 5: 679–697.

Squazzoni, F., Polhill, J.G., Edmonds, B., Edmonds, B., Ahrweiler, P., Antosz, P., Scholz, G., Chappin, E., Borit, M., Verhagen, H., Giardini, F. and Gilbert, N. (2020) 'Computational models that matter during a global pandemic outbreak: a call to action', *Journal of Artificial Societies and Social Simulation*, 23, 2. DOI: 10.18564/jasss/4298.

Stanford University Human-Centered Artificial Intelligence (2023) Artificial Intelligence Index Report 2023, https://aiindex.stanford.edu/report/ (Accessed: 18 May 2023)

Stolberg, S.G., Kaplan, T. and Robbins, R. (2021) 'Pressure mounts to lift patent protections on coronavirus vaccines', *New York Times*, 3 May. www.nytimes.com/2021/05/03/us/politics/biden-coronavirus-vaccine-patents.html?campaign_id=2&emc=edit_th_20210504&instance_id=30173&nl=todaysheadlines&regi_id=70510057&segment_id=57156&user_id=555b6d42a9884517ab975c86bed7dee1 (Accessed: 5 May 2021)

Stöppler, M.C. (2021) 'Medical definition of virus', *MedicineNet*. Reviewed 29 March. www.medicinenet.com/virus/definition.htm (Accessed: 6 September 2021)

StoryFile (2023) 'StoryFile launches Conversa AI platform for enterprise, starting Conversational Video revolution'. https://storyfile.com/storyfile-launches-conversa-ai-platform-for-enterprise-starting-conversational-video-revolution/ (Accessed: 9 May 2023)

Streeck, W. (2016) *How Will Capitalism End? Essays on a Failing System.* Verso: London and New York.

Strona, G., Bradshaw, C.J.A., Cardosa, P., Gottelli, N.J., Guillaume, F., Manca, F., Mustonen, V. and Zaman, L. (2023) 'Time-travelling pathogens and their risk to ecological communities', *PLOS Computational Biology*, 27 July. https://journals.plos.org/ploscompbiol/article?id=10.1371/journal.pcbi.1011268 (Accessed: 7 August 2023)

Stypinksa, J. (2023) AI ageism: a critical roadmap for studying age discrimination and exclusion in digitalized societies', *AI & Society*, 38, 2: 665–677. https://doi.org/10.1007/s00146-022-01553-5.

Suárez-Gonzalo, S. (2022) '"Deadbots" can speak for you after your death. Is that ethical?', *The Conversation*, 10 May. https://theconversation.com/deadbots-can-speak-for-you-after-your-death-is-that-ethical-182076 (Accessed: 10 May 2023)

Sweney, M. (2020) 'Watchdog approves use of UK phone data to help fight coronavirus', *Guardian*, 27 March. www.theguardian.com/world/2020/mar/27/watchdog-approves-use-uk-phone-data-if-helps-fight-coronavirus (Accessed: 22 August 2022)

Taylor, A. and Diamond, D. (2023) 'WHO declares covid-19 is no longer a global health emergency', *Washington Post*, 5 May. www.washingtonpost.com/world/2023/05/05/who-covid-global-health-emergency/ (Accessed: 8 May 2023)

Taylor, P. (2020) 'Susceptible, infectious, recovered', *London Review of Books*, 42, 9 (7 May): 7–10.

The BMJ (2023) 'Artificial intelligence and COVID-19'. www.bmj.com/AIcovid19 (Accessed: 3 May 2023)

The Conference Board (2023) *On the Edge: Driving Growth and Mitigating Risk Amid Extreme Volatility*. www.conference-board.org/pdfdownload. cfm?masterProductID=44769 (Accessed: 29 January 2024)

The COVID Tracking Project (2022a) Project webpage. https://covidtrack ing.com/ (Accessed: 5 September 2022)

The COVID Tracking Project (2022b) 'Thank you to our volunteers'. https://covidtracking.com/thank-you (Accessed: 5 September 2022)

The COVID-19 Host Genetics Initiative (2020) 'The COVID-19 Host Genetics Initiative, a global initiative to elucidate the role of host genetic factors in susceptibility and severity of the SARS-CoV-2 virus pandemic', *European Journal of Human Genetics*, 28: 715–718.

The Economist (2023) 'How TikTok broke social media', 21 March. www.economist.com/business/2023/03/21/how-tiktok-broke-social-media (Accessed: 29 April 2023)

The Independent Panel for Pandemic Preparedness and Response (2021) *COVID-19: Make it the Last Pandemic*. https://theindependentpanel.org/wp-content/uploads/2021/05/COVID-19-Make-it-the-Last-Pandemic_final.pdf (Accessed: 11 August 2023)

The Lancet (2023) 'Long COVID: 3 years in', 11 March. www.thelan cet.com/journals/lancet/article/PIIS0140-6736(23)00493-2/fulltext (Accessed: 6 February 2024)

The World Bank (2020) 'COVID-19 to plunge global economy into worst recession since World War II', 8 June. www.worldbank.org/en/news/press-release/2020/06/08/covid-19-to-plunge-global-economy-into-worst-recession-since-world-war-ii (Accessed: 22 January 2024)

Thompson, R.N. (2020) 'Epidemiological models are important tools for guiding COVID-19 interventions', *BMC Medicine*, 18: 152. https://doi.org/10.1186/s12916-020-01628-4.

Tolk, A. (2015) 'Learning something right from models that are wrong: epistemology of simulation', in L. Yilmaz (ed) *Concepts and Methodologies for Modeling and Simulation*. Springer: Heidelberg and New York.

Tolles, J. and Luong, T. (2020) 'Modelling epidemics with compartmental models', *JAMA*, 323, 24: 2515–2516.

Transport and Environment (2021) 'Private jets: can the super-rich supercharge zero-emission aviation?', 27 May. www.transportenvironm ent.org/discover/private-jets-can-the-super-rich-supercharge-zero-emiss ion-aviation/ (Accessed: 19 April 2023)

Triggs, A. (2021) 'The G20 should be more than just another rich-country club', *Financial Review*, 24 October. www.afr.com/world/asia/the-g20-should-be-more-than-just-another-rich-country-club-20211024-p592lc (Accessed: 8 March 2023)

Turing, A. (1950) 'Computing machinery and intelligence', *Mind*, 99, 236: 433–460.

UK Health Security Agency (2023) *UKHSA Science Strategy 2023 to 2033: Securing Health and Prosperity.* HM Government: London.

UNICEF (2020) 'Children at increased risk of harm online during global COVID-19 pandemic – UNICEF', 15 April. www.unicef.org/southafr ica/press-releases/children-increased-risk-harm-online-during-global-covid-19-pandemic-unicef (Accessed: 8 September 2021)

United Nations (2020a) 'What we do'. www.un.org/en/sections/what-we-do/index.html (Accessed: 30 June 2020)

United Nations (2020b) *United Nations Comprehensive Response to COVID-19: Saving Lives, Protecting Societies, Recovering Better.* 20 June. https://unsdg. un.org/sites/default/files/2020-06/UN-Response-to-COVID-19.pdf (Accessed: 10 August 2023)

United Nations (2020c) 'Note to Correspondents – Letter from the Secretary-General to G-20 Members', 23 March. www.un.org/sg/en/content/sg/note-correspondents/2020-03-24/note-correspondents-let ter-the-secretary-general-g-20-members (Accessed: 7 March 2023)

United Nations (2020d) *Covid-19 and Human Rights: We Are All in This Together,* April 2020. www.un.org/sites/un2.un.org/files/un_policy_brief_on_human_rights_and_covid_23_april_2020.pdf (Accessed: 30 June 2020)

UNSW (2021) 'UNSW awarded over $8m in funding for health and medical research'. www.inside.unsw.edu.au/academic-excellence/unsw-awarded-over-8m-in-funding-health-and-medical-research (Accessed: 12 September 2022)

Vaccine Confidence Project (2022a) 'About Vaccine Confidence Project™'. www.vaccineconfidence.org/about (Accessed: 13 September 2022)

Vaccine Confidence Project (2022b) 'Covid-19'. www.vaccineconfidence. org/covid-19 (Accessed: 13 September 2022)

Vaccine Confidence Project (2023) 'Projects'. www.vaccineconfidence.org/our-work/projects/ (Accessed: 5 May 2023)

Vally, H. (2022) 'Does any chance of herd immunity from COVID exist?', *newsGP,* 6 June. https://www1.racgp.org.au/newsgp/clinical/is-herd-immunity-from-covid-still-realistic (Accessed: 15 September 2022)

van Barneveld, K., Quinlan, M., Kriesler, P. et al (2020) 'The COVID-19 pandemic: lessons on building more equal and sustainable societies', *The Economic and Labour Relations Review,* First published 21 May. https://journals.sagepub.com/doi/10.1177/1035304620927107 (Accessed: 30 April 2021)

van Bergeijk, P.A.G. (2021) *Pandemic Economics.* London: Edward Elgar.

Van Beusekom, M. (2020) 'Rising vaccine wariness in some nations doesn't bode well for COVID vaccines', Centre for Infectious Disease Research and Policy, University of Minnesota, 11 September. www.cidrap.umn.edu/news-perspective/2020/09/rising-vaccine-wariness-some-nations-doesnt-bode-well-covid-vaccines (Accessed: 13 September 2022) van der Zwet, K., Barros, A.I., van Engers, T.M. and Sloot, P.M.A. (2022) 'Emergence of protests during the COVID-19 pandemic: quantitative models to explore the contributions of societal conditions', *Humanities and Social Sciences Communications*, 9: 68. https://doi.org/10.1057/s41599-022-01082-y.

van Leeuwen, H. (2020) 'The sick man of Europe', *Australian Financial Review*, 20–21 June, p 18.

Varghese, R., Patel, P., Kumar, D. and Sharma, R. (2023) 'Climate change and glacier melting: risks for unusual outbreaks?', *Journal of Travel Medicine*, 30, 4 (May 2023): taad015. https://doi.org/10.1093/jtm/taad015.

Vergano, D. (2020) 'COVID-19 might mean humanity has entered an age of pandemics, Tony Fauci warned', BuzzFeed News, 2 September, https://www.buzzfeednews.com/article/danvergano/more-coronavirus-pandemics-warning (Accessed: 15 July 2024)

Vyas, L. (2022) '"New normal" at work in a post-COVID world: work-life balance and labour markets', *Policy and Society*, 41, 1: 155–167.

Walby, S. (2015) *Crisis*. Polity: Cambridge.

Walk Free (2023) *The Global Slavery Index 2023*. https://cdn.walkfree.org/content/uploads/2023/05/17114737/Global-Slavery-Index-2023.pdf (Accessed: 24 June 2023)

Waltner-Toews, D. (2020) *On Pandemics: Deadly Diseases from Bubonic Plague to Coronavirus*. Vancouver/Berkeley: Greystone Books.

Wang, L., Zhang, Y., Wang, D., Tong, X., Liu, T., Zhang, S., Huang, J., Zhang, L., Chen, L., Fan, H. and Clarke, M. (2021) 'Artificial intelligence for COVID-19: a systematic review', *Frontiers in Medicine*, 8. DOI: 10.3389/fmed.2021.704256.

Warmbrod, K.L., Montague, M.G. and Gronvall, G.K. (2021) 'COVID-19 and the gain of function debates', *EMBO Reports*, 22, 10: e53739.

Watson, C. (2022) 'Rise of the preprint: how rapid data sharing during COVID-19 has changed science forever', *Nature Medicine*, 28: 2–4. https://doi.org/10.1038/s41591-021-01654-6.

Wenham, C. (2020) 'The UK was a global leader in preparing for pandemics. What went wrong with coronavirus?', *Guardian*, 1 May. www.theguardian.com/commentisfree/2020/may/01/uk-global-leader-pandemics-coronavirus-covid-19-crisis-britain (Accessed: 14 July 2020)

Whiting, K. and Park, H.-J. (2023) 'This is why "polycrisis" is a useful way of looking at the world right now', World Economic Forum, 7 March. www.weforum.org/agenda/2023/03/polycrisis-adam-tooze-historian-explains/ (Accessed: 25 July 2023)

Wood, R., Yannitell Reinhardt, G., RezaeeDaryakenari, B. and Windsor, L.C. (2022) 'Resisting lockdown: the influence of COVID-19 restrictions on social unrest', *International Studies Quarterly*, 66, 2: sqac015.

World Bank Group (2022) *Correcting Course*. https://openknowledge.worldb ank.org/server/api/core/bitstreams/b96b361a-a806-5567-8e8a-b1439 2e11fa0/content (Accessed: 17 April 2023)

World Economic Forum (2019) *Outbreak Readiness and Business Impact: Protecting Lives and Livelihoods across the Global Economy*. White Paper. WEF: Geneva.

World Economic Forum (2020) 'A pandemic of solidarity? This is how people are supporting one another as coronavirus spreads', 16 March. www.weforum.org/agenda/2020/03/covid-19-coronavirus-solidarity-help-pandemic/ (Accessed: 9 January 2024)

World Health Organization (2011) 'Implementation of the International Health Regulations (2005), Report of the Review Committee on the Functioning of the International Health Regulations (2005) in Relation to Pandemic (H1N1) 2009'. https://apps.who.int/gb/ebwha/pdf_files/ WHA64/A64_10-en.pdf (Accessed: 10 August 2023)

World Health Organization (2017) 'Pandemic Influenza Risk Management: A WHO Guide to Inform and Harmonize National and International Pandemic Preparedness and Response', May 2017. http://apps.who.int/iris/bitstr eam/handle/10665/259893/WHO-WHE-IHM-GIP-2017.1-eng. pdf;jsessionid=0E9957CC7B8E8EFD475ADF95F8146283?sequence=1 (Accessed: 15 February 2024)

World Health Organization (2020a) 'Origin of SARS-CoV-2', 26 March. https://apps.who.int/iris/bitstream/handle/10665/332197/WHO-2019-nCoV-FAQ-Virus_origin-2020.1-eng.pdf (Accessed: 1 August 2023)

World Health Organization (2020b) 'About WHO'. www.who.int/news-room/detail/27-05-2020-who-foundation-established-to-support-criti cal-global-health-needs (Accessed: 29 May 2020)

World Health Organization (2020c) 'Who are our stakeholders'. www.who. int/about/who-we-are/stakeholders (Accessed: 29 May 2020)

World Health Organization (2020d) 'Rolling updates on coronavirus disease (COVID-19)'. www.who.int/emergencies/diseases/novel-coronavirus-2019/events-as-they-happen (Accessed: 28 May 2020)

World Health Organization (2020e) 'COVID-19 information – SMS message library'. www.who.int/publications/i/item/covid-19-message-libr ary (Accessed: 10 August 2023)

World Health Organization (2020f) 'Director-General's remarks at the media briefing on 2019 novel coronavirus on 8 February 2020'. www.who.int/ dg/speeches/detail/director-general-s-remarks-at-the-media-briefing-on-2019-novel-coronavirus---8-february-2020 (Accessed: 3 July 2020)

World Health Organization (2020g) 'A joint campaign with the Government of the United Kingdom', 13 May. www.who.int/news-room/feature-stories/detail/countering-misinformation-about-covid-19 (Accessed: 18 July 2020)

World Health Organization (2020h) 'Impact of COVID-19 on people's livelihoods, their health and our food systems', 13 October. www.who.int/news/item/13-10-2020-impact-of-covid-19-on-people's-livelihoods-their-health-and-our-food-systems (Accessed: 22 January 2024)

World Health Organization (2021a) 'What is a pandemic?', 24 February. www.who.int/csr/disease/swineflu/frequently_asked_questions/pandemic/en/ (Accessed: 20 April 2021)

World Health Organization (2021b) 'WHO Director-General's opening remarks at the media briefing on COVID-19 – 11 March 2020'. www.who.int/director-general/speeches/detail/who-director-general-s-opening-remarks-at-the-media-briefing-on-covid-19---11-march-2020 (Accessed: 26 April 2021)

World Health Organization (2021c) *WHO Coronavirus (COVID-19) Dashboard*. https://covid19.who.int/ (Accessed: 7 July 2021)

World Health Organization (2021d) *World Health Coronavirus Disease (COVID-19) Dashboard.* (Accessed: 4 February 2021)

World Health Organization (2021e) 'Health promotion glossary of terms 2021'. www.who.int/publications/i/item/9789240038349 (Accessed: 17 July 2023)

World Health Organization (2022a) *10 Proposals to Build a Safter World Together: Strengthening the Global Architecture for Health Emergency Preparedness, Response and Resilience.* White Paper for Consultation, June 2022. https://cdn.who.int/media/docs/default-source/emergency-preparedness/2022-06-24-who-hepr-june-2022.pdf?sfvrsn=e6117d2c_3&download=true (Accessed: 13 September 2022)

World Health Organization (2022b) 'COVID-19 pandemic triggers 25% increase in prevalence of anxiety and depression worldwide', 2 March. www.who.int/news/item/02-03-2022-covid-19-pandemic-triggers-25-increase-in-prevalence-of-anxiety-and-depression-worldwide (Accessed: 2 February 2024)

World Health Organization (2023) 'WHO Director-General's opening remarks at the media briefing – 5 May 2023', 5 May. www.who.int/news-room/speeches/item/who-director-general-s-opening-remarks-at-the-media-briefing---5-may-2023 (Accessed: 8 August 2023)

World Health Organization (2024a) 'WHO COVID-19 dashboard', 21 January. https://data.who.int/dashboards/covid19/deaths?n=c (Accessed: 6 February 2024)

World Health Organization (2024b) 'Post COVID-19 condition (Long COVID)', 7 December 2022. www.who.int/europe/news-room/fact-sheets/item/post-covid-19-condition (Accessed: 8 February 2024)

World Trade Organization (2020) 'Waiver from certain provisions of the TRIPS agreement for the prevention, containment and treatment of COVID-19', Communication from India and South Africa. https://docs.wto.org/dol2fe/Pages/SS/directdoc.aspx?filename=q:/IP/C/W669.pdf&Open=True (Accessed: 29 April 2021)

World Wide Web Foundation (2020) 'There's a pandemic of online violence against women and girls', 14 July. https://webfoundation.org/2020/07/there-s-a-pandemic-of-online-violence-against-women-and-girls/ (Accessed: 8 September 2021)

Yang, C. and Tebbutt, S.J. (2023) 'Long COVID: the next public health crisis is already on its way', *The Lancet Regional Health*, 28, 100612.

Yates, K. (2021) *The Maths of Life and Death: Why Maths Is (almost) Everything*. Quercus Books: London.

Zarocostas, J. (2020a) 'How to fight an infodemic', *The Lancet*, World Report 395, 10225, P676. www.thelancet.com/journals/lancet/article/PIIS0140-6736(20)30461-X/fulltext (Accessed: 7 September 2021)

Zarocostas, J. (2020b) 'The COVID-19 infodemic', *The Lancet*, 20, 8, P875. www.thelancet.com/journals/laninf/article/PIIS1473-3099(20)30565-X/fulltext (Accessed: 7 September 2021)

Zayed, B.A., Ali, A.N., Elgebaly, A.A., Talaia, N.M., Hamed, M. and Mansour, F.R. (2023) 'Smartphone-based point-of-care testing of the SARS-CoV-2: a systematic review', *Scientific African*, 21: e)1757.

Zhang, W.Q., Montayre, J., Ho, M-H., Yuan, F. and Chang, H-C. (2021) 'The COVID-19 pandemic: narratives or front-line nurses from Wuhan, China', *Nursing and Health Sciences*, 24: 304–311.

Zuboff, S. (2019) *The Age of Surveillance Capitalism: The Fight for a Human Future At the New Frontier of Power*. Profile Books: London.

Zurcher, A. (2020) 'Coronavirus response: things the US has got right – and got wrong', *BBC News*, 13 May. www.bbc.com/news/world-us-canada-52579200 (Accessed: 25 June 2020)

Index

A

advertising 95
Affectiva 95
affective computing 95
affective responses *see*
    emotion-based responses
Agamben, Giorgio 29–30
age inequalities 14
AI *see* artificial intelligence
air travel 84–85
aleatory uncertainty 53
algorithms 94
Anderson, Warwick 53, 60
Anirudh, A. 65
Arendt, H. 33
artificial intelligence (AI) 9, 57–58, 90–91,
    98–101, 123, 127–128
Artificial Intelligence Index Report 100
AstraZeneca 104
Aucante, Y. 10
Australia
    economic measures 75
    Federal and state policy 15, 111
    healthcare 102–103
    influence of modelling 62, 69–70
    lockdowns 117–118
    wealth disparities 84, 102–103
authoritarian regimes 35
Avian flu (H5N1 influenza) 53, 112

B

Baker, R.E. 124
Becker, Gay 31–32, 33–34
Bernoulli, Daniel 54
Bessen, J. 90
Bhattacharya, Jay 64
big data 21, 87
    see also digital technologies
Big Tech companies 90, 93–94
bio-digital citizenship 9
Black Swan events 29
blame 17–18, 119
Bloomberg 58–59
Box, George 51–52

Brainard, Jeffrey 4
bubonic plague 109
Budd, William 54

C

Caduff, Carlo 61
call data records (CDR) 88
Capano, G. 43
capitalism 35, 94–95, 123
Capoccia, G. 36
Carstens, Agustin 75–76
cascading effects 13, 26, 48
Centers for Disease Control and Prevention
    (CDC) 47, 55, 109
chaos, concepts of 31–32
chaos theory 31
China 10, 16–18, 35
civil unrest and resistance 47, 57, 70, 75,
    115–116
climate change 48–49, 122–123
climate lockdown 45–46
Coburn, B.J. 54–55
Connell, R. 128
cooperation/collaboration 54, 68–69, 78–79,
    87, 102
    see also data sharing
coordination *see* solidarity
Covid Resilience Ranking 58–59
Covid Tracking Project 66
COVID-19 Genomics UK Consortium
    (COG-UK) 68
COVID-19 Host Genetics Initiative 68–69
COVID-19 pandemic
    affective responses to 15–16, 33–34,
        118–119
    death rates 69–70, 109
    declaration as crisis 36–42
    experiences of crisis 27–28
    framing of *see* frames and framing
    impacts *see* social and economic impacts of
        COVID-19
    and inequality *see* inequalities
    modelling *see* models and modelling
    origins and responsibility 17–18

policy responses *see* policy responses to COVID-19 pandemic
research 3–6, 66–67, 124–125
and technology *see* digital technologies
Credit Suisse Research Institute 81–82
crises
and appeals to 'health' 30–31
characteristics 13–18
as constructed 13
COVID-19 pandemic framing 7–8, 36–42, 108–110
disruption related to 14–15, 27–28, 31–32
emotional dimensions 15–17, 32–33, 118–119
and identity questioning 33–34
impacts *see* social and economic impacts of COVID-19
as opportunity for change 124
polycrisis 124
sociological analyses of 26–27
in systemic context 34–35
and use of state powers 28–30
crisis messaging 39–40, 71–74
critical junctures 36
Cummings, Dominic 47
cybercrime 89

D

data, impact during pandemic 21–22, 87
data brokers 105
data centres 105
data infrastructure 104–105
data security 89–90
data-sharing 55, 66–67, 87–91
death rates 69–70, 109
Defensor, G. 105
developing countries 73, 77–78, 120
digital technologies
accelerated by COVID-19 pandemic 93–94, 99, 101–102, 104
AI 9, 57–58, 90–91, 98–101, 123, 127–128
data infrastructure growth 104–105
data-sharing systems 87–91
health innovations 101–104
impacts on everyday life 91–93
and inequality 82, 105
and national science strategies 105–107
online dashboards 22, 66
role in COVID-19 pandemic 9–11, 21–22, 34, 86–87
tech companies 90, 93–94
videofication 94–98
Dingwall, R. 48
discrimination 118
disease
fear of 16–17, 30–31, 72
modelling applied to 53–55
Disinfo Lab 46
disruption 14–15, 27–28, 31–32
self-questioning following 33–34

Doherty, Peter 62, 121
Drake, John 68
Dubos, R. 30
Durkheim, Emile 32

E

Ebola 55
economic impacts *see* social and economic impacts of COVID-19
economic stimulus measures 74–76
economy, link with health 123
emotion-based responses 15–17, 32–34, 118–119
see also fear
emotion research, and digitization 95–96
employment
hybrid/flexible working 91, 92
loss of 14, 75, 83, 92
platform work 104–105
and self-questioning 34
support packages 74–75
technological impacts 91–92, 99
Englemann, L. 60
Enserink, M. 63–64
environmental harm 84–85
see also climate change
epidemiology 125
epistemic uncertainty 53
EPIWATCH 57–58
eugenics 30
European Union Disinfo Lab 46
experts 21, 60, 62, 126–127
disagreement between 62–63, 114–115, 127

F

Facebook 45–46
facial recognition 88
factual information 46, 114–115
false hope 119
Farr, William 54
Fassin, Didier 120
Fauci, Anthony 121
fear 16–17, 30–31, 61–62, 72, 119
Ferguson, Neil 53, 59–60, 62–63
financial crisis 48
flexible/hybrid working 91, 92
Fourcade, Marion 120
frames and framing
concepts 6, 37
consequences 115
contested 8, 126–127
of pandemic as crisis 7–8, 36–42, 108–110
risks of AI tools 128
role of modelling *see* models and modelling
role of news media 8
as 'state of exception' 7–8, 28–30
Frandsen, Mikkel 121
freedom, reasons for restricting 7, 30, 117–118

G

G20 countries 39–40, 71–73, 76–79
gender inequalities 14, 75
generative AI 100–101
genetics 30
genomic sequencing 67–69
Ghebreyesus, Tedros Adhanom 44, 89, 110, 116
Giesecke, Johan 10
Gilbert, N. 122
Glass, James 30
global context
  impacts of COVID-19 pandemic 15
  inequality 76–79, 80–81, 82, 120–121
  solidarity 15, 38–39, 72–73, 74, 114
  see also United Nations; World Bank; World Health Organization
global economy 73
global governance reform 110–111
  constraints on WHO 113–114
  and emotion-based responses 118–119
  expectations of science and technology 114–116
  and historical analogies 111–113
  media broadening of debate 121–124
  and uncertainty 116–118
  and unequal policy impacts 119–121
Goffman, Erving 6
Google 101
Gostin, L.O. 114
governance see global governance reform; risk governance approach
government/state
  extension of powers 7–8, 28–30, 39, 88, 117–118
  historic appeals to 'health' 30–31
  see also policy responses to COVID-19 pandemic; politics
Great Barrington Declaration 64–65
GSMA (Global System for Mobile Communications) 87
Gupta, Sunetra 63, 64
Guterres, António 8, 72

H

H1N1 influenza (Swine flu) 54, 112
H5N1 influenza (Avian flu) 53, 112
Habermas, Jürgen 35
Hacking, Ian 13
Harvey, David 94–95, 96
health
  appeals to 30–31
  link with economy 123
  see also global governance reform
health innovations (digital) 101–104
health security 106
Health Security Agency 106
healthcare systems 61–62, 101–102
herd immunity 10–11, 59–60, 111

hero metaphors 21, 40
historical analogies 111–113
historical appeals 30–31
home-based working 91, 92
hope 16, 119
Horton, Richard 126–127
housing costs 83
human trafficking 83
hybrid work model 91, 92

I

identity, questioning of 33–34
Imperial College model 59–63
The Independent Panel for Pandemic Preparedness and Response 79–80, 113
inequalities
  global 76–79, 80–81, 82, 120–121
  role of science 106
  and technology 82, 105
  unequal impacts of COVID-19 pandemic 14–15, 29, 64, 75, 76–79, 80–85, 106, 119–121
  vaccine nationalism 78–79, 80–81, 106, 114
  wealth disparities 29, 81–85, 119–121
inflation 75–76
influenza pandemics 54, 109, 112, 113
infodemic 44–48, 89
infographics 21, 22
informal economy 83
inscription devices 20
intellectual property rights 78–79
interest rates 75, 76, 77, 83
internal migration 92–93

J

Johnson, Boris 40–41, 47

K

Kelemen, R.D. 36
Kiran, H. 105
Kirby Institute 57–58
Klein, Naomi 28
Kucharski, Adam 65–66
Kulldorff, Martin 64
Kupferschmidt, K. 63–64

L

'lab leak' hypothesis 17–18
Lancaster, Kari 56
language, in crisis messaging 39, 40, 44
Latour, Bruno 20, 123
Leach, M. 19
liberal democracies 35
lockdowns
  harmful impact 64–65, 83, 117–118
  models justifying 60
  national approaches 10, 36, 42, 110–111, 117–118

public responses 70
resistance to 47, 70, 75
and videofication 94
Long COVID 109–110
Lovejoy, Kris 89–90
Lupton, Deborah 125
Lyon, David 87

M

McKinsey and Company 93–94
media 8, 121–124
social media 44–47, 95–98
mental illness 103
Metzler, H. 96
Microsoft Teams 94, 98
military metaphors 39, 40, 44
misinformation 44–47
models and modelling
concept 20, 50, 51–53
Covid Resilience Ranking 58–59
EPIWATCH 57–58
and fear 61–62
history of epidemiological 53–56
influential modellers and models 62–64
limitations and concerns 51–53, 55, 60–61,
62, 63–65, 69
public representations of 65–70
and risk governance 20–22,
56–57
SIR model 20, 54, 61–62, 65
as technologies of framing 52
UK debates during COVID-19
pandemic 59–61
of vaccine confidence 57
modern slavery 83–84
Mooney, C. 37

N

national science strategies 105–107
Nazi Germany 30–31
news media 8, 121
Nextstrain 67–68
Nisbet, M.C. 37
Norberg, J. 33

O

Oberkampf, W.L. 53
older people 14
Omicron variant 69–70
online dashboards 22, 66
order and chaos 31–32
Organisation for Economic Co-operation and
Development (OECD) 87–88
Oxford model 63

P

pandemics
crises as opportunity for change 124
danger of historical analogies 111–113

death rates 109
definition 111, 112–113
future 121–122
preparedness 41–42, 79–80, 89
as socio-political phenomena 2
see also COVID-19 pandemic
Panovska-Griffiths, J. 63
PCR (polymerase chain reaction)
tests 19, 103
Peckham, Robert 111
Pennington, Hugh 106–107
Pfizer 104
pharmaceutical companies 78–79, 81
platform work 104–105
policy responses to COVID-19 pandemic
critical evaluations of 79–81
economic stimulus 72–73, 74–76
extension of state powers 7–8, 28–30, 88,
117–118
harmful impacts 2–3, 64–65, 115–116,
117–118, 119
health and economy tensions 123
inequitable impacts 64, 75, 76–79, 80–85,
119–121
initial US and UK responses 36, 40, 41–42,
110–111
national variations 9–11, 15, 76–77, 80,
110–111
risk governance approach 19–22, 43
role of modelling 56–57, 59–61
UN and WHO calls for solidarity 38–39,
72–73, 74
UN and WHO crisis messaging 39–40,
71–74
see also global governance reform
politics, and science 10, 11–12
polycrisis 124
poverty 14, 81, 83
predictive models 51
preparedness 28, 41–42, 79–80, 89
preprints 66–67
privacy 88
private jets 84–85
proprietary software 90
protests/civil unrest 47, 70, 75,
115–116
publication of research 4–5, 66–67

R

rapid antigen tests (RATs) 19, 104, 126
relocation 92
Renn, Ortwin 42, 43
reproduction rate (R_0) 53
research 3–6, 66–67, 124–125
resistance/civil unrest 47, 70, 75, 115–116
responses see emotion-based responses;
policy responses to COVID-19 pandemic
responsibility/blame 17–18, 119
Rhodes, Tim 56, 65

rights, suspension of 7, 28, 29, 88, 117–118
risk governance approach 19–22, 25, 42–44, 45, 56–57, 112, 123

S

science-based information 46, 47
science and technology
  central role 7
  cooperation/collaboration 68–69
  national strategies 105–107
  as political 10, 11–12
  tempering expectations of 114–116
  see also digital technologies; experts
security (data) 89–90
security (health) 106
self-testing 19, 104, 126
sexual exploitation 84
shadow economy 83
Sharma, A. 104
SIR (Susceptible Infectious-Recovered)
  model 20, 54, 61–62, 65
slavery 83–84
Snow, Jon 54
social distancing 47
social and economic impacts of COVID-19
  acknowledging uncertainty 117–118
  cascading effects 13, 48
  crisis messaging 72, 73–74
  critical evaluations 79–81
  economic stimulus measures 72–73, 74–76
  G20 responses 76–79
  Long COVID 109–110
  resilience rankings 58–59
  unequal impacts 14–15, 29, 64, 75, 78, 81–85, 106, 119–121
  wealth disparities 29, 81–85, 119–121
social media 44–47, 95–98
social renewal 2, 124–128
software 90, 105
solidarity (global) 15, 38–39, 72–73, 74, 114
state *see* government/state
'state of exception' 7–8, 28–30, 88
stock markets 29
surveillance
  and digitization 87
  EPIWATCH 57–58
  global infrastructure 21–22
  role of AI 98–99
  tracking 66, 87, 88
Susceptible-Infectious-Recovered (SIR)
  model 20, 54, 61–62, 65
Sweden 10–11, 111
Swine flu (H1N1 influenza) 54, 112

T

tech companies 90, 93–94
technological 'arms race' 90

technology *see* digital technologies;
  science and technology
Tegnell, Anders 10
telehealth 102–103
testing 19, 103–104, 126
Thompson, R.N. 62
TikTok 94, 96–98, 101
time frames (research) 4–5, 124–125
time-space compression 94–95, 96
Tolk, Andreas 52
totalitarianism 33
tracking 66, 87, 88
Triggs, A. 77, 78
Turing, Alan 99
Twitter 45–46, 96

U

uncertainty 19, 43, 52–53, 116–118
unemployment 14, 75, 83, 92
United Kingdom (UK)
  debates on modelling 59–61
  initial pandemic response 36, 40, 41–42, 110–111
  preparedness 41–42
  response to misinformation 47
United Nations (UN) 37, 38–40, 71–74
United States (US)
  end of COVID-19 pandemic 12
  initial pandemic response 36, 40, 42
  response to misinformation 47
utilitarianism 70, 118

V

Vaccine Confidence Project (VCP) 57, 96
vaccine development 104, 106
vaccine hesitancy 57, 75
vaccine nationalism 78–79, 80–81, 106, 114
vaccine sharing 54, 78–79
videofication 94–98
viruses, fear and danger of 16–17
visual depictions 21, 22

W

Walby, Sylvia 13, 26–27, 38, 39, 44, 48
Walk Free 83–84
Waltner-Toews, D. 112
Watson, C. 67
wealth disparities 29, 81–85, 119–121
women, impact of pandemic on 14, 75
Woolgar, Steve 20
work *see* employment
World Bank 28, 73, 82–83
World Health Organization (WHO)
  call for solidarity 38–39, 74
  constraints on 81, 113–114
  COVID-19 framing and reframing 38
  data and communication 21–22, 40, 88–89

end of COVID-19 pandemic 12
function 37
lessons learned 116
pandemic definition and response 111–113
on pandemic impact 110
response to misinformation 46
Wuhan markets 17–18

Y
YouTube 45–46

Z
Zoom 94, 98, 101
zoonotic transfer 18
Zuboff, Shoshana 56